MEASURING MASS

FROM POSITIVE RAYS TO PROTEINS

MEASURING MASS

FROM POSITIVE RAYS TO PROTEINS

Edited by Michael A. Grayson
for the American Society for Mass Spectrometry

Chemical Heritage Press
Philadelphia

Printed in the United States of America.

For information about CHF and its publications contact

Chemical Heritage Foundation
315 Chestnut Street
Philadelphia, PA 19106-2702, USA
Fax: (215) 925-1954
www.chemheritage.org

For information about ASMS and its publications contact

American Society for Mass Spectrometry
Building I
2019 Galisteo Street
Sante Fe, NM 87505
www.asms.org

Book designed by Mark Willie, Willie•Fetchko Graphic Design

Printed by Alcom Printing Group

Library of Congress Cataloging-in-Publication Data
Measuring mass : from positive rays to proteins / edited by Michael A. Grayson.
p. cm.
Includes bibliographical references and index.
    ISBN 0-941901-31-9 (alk. paper)
    1. Mass spectrometry—History. I. Grayson, Michael A.
    QD96.M3 M43 2002
    543'.0873'09—dc21
                                2001007646

This book is printed on acid-free paper.

Contents

Foreword

Mass spectrometry enjoys a rich heritage dating back more than a hundred years. Although rooted in physics, today it has much broader and more powerful capabilities for scientists to use to uncover the complexities of elemental and molecular processes in nature. Mass spectrometry has been an integral part of historically significant events in the past century, including discoveries and vital measurements in biochemistry, chemistry, and physics as well as in human history in both war and peace. It has appeared at the forefront of science throughout the century and indeed may yet have its greatest role to play in coming years in the quest to improve human health and combat disease.

The development of mass spectrometric instrumentation closely followed other exciting innovations in science over the decades. Mass spectrometrists eagerly embraced the new ideas and swiftly incorporated them into their apparatus. Today, ultra-high-speed electronics and computers are vital to this instrumentation and are allowing the innate capabilities of the discipline to be expanded to their fullest extent.

Above all, mass spectrometry is able to provide detailed information quickly and accurately about chemical structure and molecular dynamics. In doing so, it has led to fundamental discoveries in diverse fields of science and other human endeavors. Consider, for example, the elucidation of the molecular dynamics of living cells, which involve the balanced interrelationship of anabolic and catabolic processes. This landmark discovery in biochemistry was made possible through the analytical capabilities of isotope ratio mass spectrometry during the 1930s and 1940s. The problem-solving ability of mass spectrometry continues today in nearly every discipline involving elemental and molecular analysis.

Mass spectrometry instrumentation has developed in a unique way. Instruments that have achieved phenomenal increases in resolution, speed, and sensitivity have generally come in smaller and less expensive packages. These instruments can be easily mastered by users unskilled in the underlying principles, allowing for the proliferation of mass spectrometry in nearly all fields of science. The reliability and reproducibility of mass spectrometers have achieved a remarkable level, further supporting the migration of this technology to other fields.

The mass spectrometry community has every reason to be proud of its heritage and the future of mass spectrometry. Committee E-14 of the American Society for Testing and Materials was the initial organization of mass spectrometrists in the United States. In later years, this group evolved into the American Society for Mass Spectrometry (ASMS).

ASMS is proud to celebrate the legacy and future of mass spectrometry in science by supporting the publication of *Measuring Mass: From Positive Rays to Proteins* on the fiftieth anniversary of the Conference on Mass Spectrometry and Allied Topics. This volume can provide only a glimpse of the history and accomplishments of those who have pioneered and used mass spectrometry in innovative ways. We congratulate the worldwide community of scientists in both fundamental and applied research for their contribution to the success of this versatile and vitally important technology.

Richard M. Caprioli
President, ASMS
9 January 2002

Acknowledgments

The efforts of many individuals were crucial to the creation of this book. In particular, special thanks go to Robert Cotter, who, while president of the ASMS, initiated and supported this project. Richard M. Caprioli, current president of the ASMS, continued that enthusiastic support to see the project through to a successful completion. David Brock, historian and program manager of the Beckman Center, was a tower of strength throughout the entire project. His diplomatic skills, critical eye, and attention to detail kept the writers on track. The final product has benefited enormously from his contributions. A debt of gratitude must also be paid to Leo Slater, who helped initiate the project at CHF.

Many others in the ASMS and the greater scientific community gave their time and expertise in various phases of the project. Thanks go to Klaus Biemann, Bill Budde, Donald Chace, R. Graham Cooks, Jan Crowley, Leslie Ettre, Bryan Finkle, P. Jane Gale, Ron Hites, Joseph Loo, Karleugen Habfast, Keith Jennings, Dwight Matthews, Konrad Mauersberger, Birenda Pramanik, Kevin Yarashevski, and Jehuda Yinon. Thanks also to all of those who contributed to the Web-page timeline and many others who made contributions. This book was a labor of love for many dedicated individuals, and the quality of the work is their deserved reward.

At CHF, historian Thomas Lassman identified source materials in the published scientific and trade journal literature. Christina Dunbar-Hester and Suzanne Warren spent countless hours collecting images and photographs, while Kristen Graff patiently filled interlibrary loan requests for dozens of articles and books on the history of mass spectrometry. Laura Wukovitz provided valuable reference assistance. Mark Michalovic helped assemble and edit the entries for the historical markers timeline in the text. Their efforts are much appreciated.

When the manuscript reached CHF's publications area, Shelley Wilks Geehr shepherded the rough manuscript through the production process to the final beautifully scripted color layout. Her experience and advice have been invaluable during the project. Patricia Wieland provided expert editing assistance, Suzanne Warren provided necessary administrative support, and freelancer Martha Witte lent her expertise by obtaining prints and permissions for the images that enhance this publication.

Contributors

Editor

Michael A. Grayson
Assistant Director of the Mass
 Spectrometry Research Resource
Washington University in St. Louis
St Louis, Missouri

Contributors

J. Thomas Brenna
Professor of Nutrition and of Chemistry
 and Chemical Biology
Cornell University
Ithaca, New York

Kenneth L. Busch
Office of Sponsored Programs
Kennesaw State University
Kennesaw, Georgia

Richard M. Caprioli
Stanley Cohen Professor of Biochemistry
Professor of Chemistry and Pharmacology
Director, Mass Spectrometry Research Center
Vanderbilt University
Nashville, Tennessee

Robert J. Cotter
Professor of Pharmacology and Molecular Sciences
Johns Hopkins University School of Medicine
Baltimore, Maryland

Ronald D. Grigsby
Principal Chemist
TRW Petroleum Technologies
Bartlesville, Oklahoma

Charles M. Judson
Emeritus, Department of Chemistry
University of Kansas
Lawrence, Kansas

Ragu Ramanathan
Principal Scientist
Shering-Plough Research Institute
Drug Metabolism and Pharmacokinetics
Kenilworth, New Jersey

Gary Siuzdak
Director of The Scripps Center for
 Mass Spectrometry
Professor of Molecular Biology
The Scripps Research Institute
La Jolla, California

Michael S. Story
Vice President for Mass Spectrometry
ThermoFinnigan Corporation
San Jose, California

John J. Thomas
Staff Scientist in Protein Therapy Research
Genzyme Corporation
Framingham, Massachusetts

Ross C. Willoughby
President
Chem-Space Associates
Pittsburgh, Pennsylvania

Alfred L. Yergey
Section Chief
National Institute of Child Health and
 Human Development
National Institutes of Health
Bethesda, Maryland

Molecules having the same mass numbers but differing in weight by an amount determined only by the difference in binding energies of the nuclear particles can be clearly resolved. . . . Extension of the use of the instrument [Nier–Johnson double focusing mass spectrometer] to the resolution of heavy hydrocarbons should prove fruitful.

Alfred O. Nier
"Determination of Isotopic Masses and Abundances by Mass Spectrometry," *Science* 121 (1955), 740.

Chapter **1**
Origins

Above: Cambridge University physicist Ernest Rutherford discovered the atomic nucleus in 1911.

Middle: John Dalton.

Right: Dmitri Mendeleev.

The nineteenth and early twentieth centuries were marked by major intellectual transformations in the physical sciences. New discoveries—especially radioactivity, X-rays, the electron, and the atomic nucleus—had forced physicists and chemists to call into question their understanding of the physical world. New experimental methods were developed and new theories proposed to explain controversial observations of the composition and behavior of matter.

In 1803 John Dalton proposed a new atomic theory to explain the known properties of matter. Unlike his predecessors Isaac Newton and Robert Boyle, Dalton did not believe that matter was composed of a single, homogeneous substance. Dalton's atomic theory rested on four critical assumptions: all matter is composed of elements consisting of solid, indivisible atoms; the elements are indestructible; the number of different types of atoms correspond exactly to the number of chemical elements; and each element has a fixed, measurable property—relative atomic weight. As a consequence of Dalton's theory, when atoms combine to form compounds, a given weight of one element reacts with a fixed weight of one or more other elements in ratios of small whole numbers. Dalton's conceptual innovations were soon followed by William Prout's theory of integral atomic weights (1815) and by Dmitri Mendeleev's discovery of the periodic law of the elements (1869).

As the nineteenth century drew to a close, however, Dalton's ideas concerning the behavior and structure of atoms began to break down in the face of new discoveries in physics. Experimental studies of electrical discharges in gases gave strong evidence that atoms were not indivisible particles, as Dalton and his contemporaries believed. Corresponding developments in the study of radioactivity also weakened the foundations of this theory.

Left: Francis Aston and his third mass spectrograph, 1937.

MASS SPECTROMETRY

1886 Eugen Goldstein observes canal rays, positive rays of electricity.

1897 J. J. Thomson discovers the electron.

HISTORY

 **1895** X-rays are discovered by Wilhelm Roentgen.

 Westinghouse Electric installs the first alternating current generators for the production of electricity at Niagara Falls, New York.

 1898 Marie and Pierre Curie discover radium.

In 1886 the German physicist Eugen Goldstein identified a new type of radiation, which he called *Kanalstrahlen,* or "canal rays." Goldstein had produced the rays by passing an electrical discharge through a gas at low pressure in a glass tube equipped with a perforated cathode. In conventional discharge tubes, negatively charged particles, known as "cathode rays," were observed to stream from the cathode to the anode. Goldstein's rays, however, traveled in the opposite direction. He concluded that canal rays were composed of positively charged particles. Unlike the negative particles in cathode rays, *Kanalstrahlen* gave rise to a glow discharge, the color of which varied in accordance with the nature of the gas in the tube. For example, the color of the discharge in air was yellow; in hydrogen, rose; and in carbon dioxide, greenish-gray. Goldstein also demonstrated that canal rays were unaffected by weak magnetic fields placed near the discharge tube.

In 1898 physicist Wilhelm Wien showed that canal rays could be deflected by strong electric and magnetic fields. When Wien used superimposed parallel electric and magnetic fields for ion deflection, he found that particles with different charge-to-mass ratios (e/m) followed different parabolic curves. On the basis of this evidence Wien concluded that the e/m values of canal rays depended on the nature of the gas in the tube.

Prompted by interest in Wien's experimental results, J. J. Thomson, a physicist, began his own studies of Goldstein's *Kanalstrahlen* in 1905 in the Cavendish Laboratory at Cambridge University. The parabolas obtained by Wien were blurred, which Thomson believed was an

Top: Eugen Goldstein.

Bottom: Wilhelm Wien.

Of Mass and Charge

When physicists began exploring the world of charged particles, they adopted the term *charge-to-mass ratio,* or e/m, to give scientific and descriptive units to the measurements taken from their early instruments. This terminology reflected their interest in the nature of the charged particles they were studying. In the case of mass spectrometry, however, the charge-to-mass ratio is really not useful in the interpretation of a mass spectrum. Analytical chemists are only interested in the mass. Therefore, they adopted the term *mass-to-charge ratio,* or m/z.

For years mass spectrometrists referred to the peaks in a mass spectrum as the "mass" of the ion fragments represented. This quantity for mass was referred to as "m/e," where e represented the charge on an electron. While convenient, this method is impre-

cise, since the use of e implied that all ions were singly charged. The peaks in a mass spectrum really represent the ion current measured by the instrument at a given value of m/z. A peak at $m/z = 22$, for example, could be from a singly charged ion of the ^{22}Ne isotope (22/1), or it could be from a carbon dioxide ion with two charges (44/2). Referring to the peak as having a mass of 22 implied that it has only one charge and thus would have to be from the ^{22}Ne isotope. But to refer to the peak at $m/z = 22$ does not provide any implied or understood information about the nature of the peak itself. This distinction is important, since multiply charged ions appear in the mass spectra of many compounds. The symbol m/z is used to represent the mass-to-charge ratio with the understanding that z can represent any number of charges.

MASS SPECTROMETRY	1898	1899
	Wilhelm Wien studies canal rays, deflecting them with electric and magnetic fields.	A. A. Campbell Swinton conducts experiments on the "reflection" of cathode rays with magnetic fields.

HISTORY	1898	1899
	The Spanish-American War ends in a U.S. victory and in U.S. possession of Puerto Rico, Guam, and the Philippines.	Guglielmo Marconi sends the first radio transmission across the English Channel.

The Mass Spectrum: Thomson's Parabola Mass Spectrograph

A mass spectrum from J. J. Thomson's early instrument appears strange to the modern eye. It consists of a series of parabolic lines radiating above and below the middle of a photographic plate. The spectrum looks this way because Thomson's instrument used parallel and coterminal electric and magnetic fields. Ion motion was affected by both the magnetic and electric fields simultaneously.

Ions were created in a gas discharge apparatus and accelerated through a fine-bore opening in the cathode of the tube, providing collimation of the ion beam. Once the ions passed through the combined fields, they were separated simultaneously by mass and energy. Since mass is an integral quantity, we see a series of parabolas, one for each mass. The displacement along the parabola is a measure of the ion energy.

Since the ions were created in a gas discharge, they possessed a range of energies. Because the electric field of Thomson's mass spectrograph deflected energetic ions only slightly, most of them struck the photographic plate closest to the vertex of the parabola. Furthermore, the intensity of the parabola in the region of high energy is greatest, indicating that most of the ions formed in the gas discharge have a high energy. Consequently, the less energetic ions formed in the gas discharge lie on the parabola farther from the vertex. The intensity of the parabolic line in this region indicates that there are fewer low-energy ions in the gas discharge.

An interesting feature of this instrument is that negative ion data is recorded simultaneously with positive ion data. The parabolas on the upper half of the photographic plate are from positive ions, while the parabolas on the lower half are from negative ions. In actual operation only two of the four quadrants of the photographic plate will record data during the experiment. The upper and lower left-hand quadrants record data from positive and negative ions, respectively. After recording spectra in these quadrants, the polarity of the magnetic and electric fields is reversed, and spectra are then recorded in the other two quadrants. This was done to improve the ease and accuracy of data analysis of the photographic plate.

Finally, the equations of ion motion for the parabolic mass spectrometer are expressed in terms of "e/m," charge-to-mass ratio. Data from early instruments was reported in the scientific literature in these terms. Only later did physicists and chemists adopt the more physically meaningful quantity "m/z" to denote the lines in a mass spectrum.

Reference

R. W. Kiser. *Introduction to Mass Spectrometry and Its Applications.* Englewood Cliffs, N.J.: Prentice Hall, 1965.

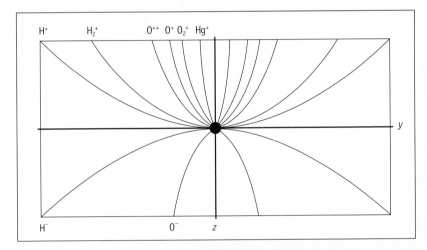

Figure 1. *Hypothetical photographic record of a parabola mass spectrograph. Reproduced from R. W. Kiser,* Introduction to Mass Spectrometry and Its Applications *(Englewood Cliffs, N.J.: Prentice Hall, 1965).*

1900

The Boer War rages in South Africa between British colonial forces and the descendants of Dutch settlers known as Boers.

Sigmund Freud publishes *The Interpretation of Dreams*.

Chinese peasants and Imperial armies join forces in the Boxer Rebellion against the influence of foreign powers. The rebellion is quelled by European and American forces.

J. J. Thomson.

Right: The Cavendish Laboratory at Cambridge University was the world center for experimental physics at the beginning of the twentieth century.

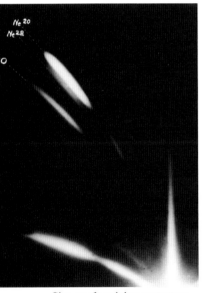

Close-up of parabola mass spectrum of neon.

indication of variability in either the atomic weights of the constituent particles or the magnitudes of their electric charges. Further studies led Thomson to attribute the spread in Wien's parabolas to collisions between the positive rays streaming through the cathode and residual gas molecules present in the discharge tube. After constructing a new tube that operated at lower pressures, Thomson eliminated the collision problem and obtained sharp parabolas for hydrogen at positions corresponding to e/m values of H^+ and H_2^+. He further observed that oxygen gave three separate curves corresponding to O^+, O^{++}, and O_2^+ and that the introduction of polyatomic molecules, such as $COCl_2$, into the tube yielded parabolas corresponding to C^+, O^+, Cl^+, CO^+, Cl_2^+, and $COCl_2^+$. Thomson also recorded parabolas for ions generated from hydrocarbon molecules.

The photographic plate Thomson used in his mass spectrograph to record the parabolas in these early experiments showed greater sensitivity to light ions than to heavy ions. This limitation introduced errors in recording the relative numbers of ions of different atomic compositions. Thomson contrived an ingenious way to circumvent this problem. In place of the photographic plate he used a metal plate in which a parabolic slit had been cut. By varying the strength of the magnetic field, he was able to focus each ion beam onto the parabolic slit and measure the current of the ions as they passed through the slit onto a second metal plate connected to an electroscope. By plotting the resulting ion current as he varied the strength of the magnetic field, Thomson was able to generate a graph of ion abundance versus mass. The most intense peaks corresponded to the most abundant ions. Thomson had produced a mass spectrum in what became known as the first mass spectrometer.

The Discovery of Isotopes of the Stable Elements

When Thomson used neon in the discharge tube, he noticed an anomaly that could not be explained by existing theory. Two parabolas appeared, one corresponding to atomic weight 20 and the other to 22. William Ramsay had shown in 1898 that the atomic weight of neon was 20.2, which made the second line on Thomson's plate especially troubling. Several possible solutions were proposed. The ions at 22 might correspond to a hydride of neon (NeH_2), or perhaps they signified the presence of doubly charged carbon dioxide. Thomson himself believed that the ions could be explained by the presence of another element having an atomic weight close to the value for neon.

MASS SPECTROMETRY		1901
		German physicist W. Kaufman creates the predecessor to the parabola mass spectrograph to study cathode rays.

HISTORY	1900		1901
	Max Planck introduces the quantum theory of black body radiation.	The Kodak Brownie camera goes on sale, bringing photography to the masses.	President William McKinley is assassinated.

Spectrometers and Spectrographs: How Ions Are Detected

In a mass spectrometer the ion image is formed by a slit at the entrance of the analyzer and focused onto a slit in front of the detector. The spectrum is recorded by scanning electric or magnetic fields, or both, of the spectrometer in a controlled fashion, thus causing ions of different mass to arrive at the detector slit at different times. The detector measures and amplifies the variations in ion current during a scan, creating a voltage signal that is subsequently amplified further before being recorded. The spectrum then appears as a series of peaks, the intensity of each being proportional to the ion current at a particular mass.

In the mass spectrograph the ion image is formed by a slit, just as in the mass spectrometer, but the ions are focused onto a plane, and the detector is a photographic plate placed at that focal plane. Both the electric and magnetic sector fields are held at constant values so that the desired range of masses will fit across the breadth of the photoplate. The spectrum on the photoplate looks like a series of lines, essentially multiple images of the source slit, one for each mass in the spectrum, spread across the photoplate. The darkness of a line is a measure of the ion current for that mass accumulated during the exposure of the photoplate. Special equipment is required to "read" the lines on the photoplate and produce a mass spectrum similar to that of the mass spectrometer.

The mass spectrograph makes use of an integrating detector; most of the ions that leave the source are recorded somewhere on the photoplate. This is not the case for the mass spectrometer, which makes use of an electric detector. Only ions of the mass focused on the detector are recorded. Ions of all other masses are discarded.

Francis Aston.

A satisfactory explanation of the extra line in the mass spectrum of neon remained elusive until Thomson's new assistant, Francis Aston, extended the recently developed concept of radioactive isotopes to stable elements in the periodic table. Aston, born in 1877, had studied chemistry and physics at Birmingham University, where he had demonstrated a keen aptitude for the design and construction of experimental apparatus. Aston became a skilled glassblower, and before moving to Cambridge in 1909, he built several types of vacuum pumps and small discharge tubes. He drew heavily on this expertise in his early attempts to explain the neon spectrum in Thomson's mass spectrograph.

At first Aston tried to identify the origin of the e/m 22 parabola by using conventional chemical separations. Fractional distillation, fractional diffusion, differential absorption, and even the use of a very sensitive quartz microbalance yielded no more than partial separation of the gas. Further research along these lines was interrupted by the onset of World War I.

After the war Aston attacked the neon problem from an entirely different angle. Instead of returning to the established chemical separation methods to explain Thomson's experimental results with neon, he extended the concept of isotopes from the radioactive to the stable elements. Frederick Soddy and his associates at the University of Glasgow had shown earlier that the natural decay of radioactive elements, such as thorium and uranium, produced other substances with chemical properties identical to those of known elements, such

Australia achieves nationhood with the creation of the Commonwealth of Australia from the six British colonies that occupied the continent.

The Nobel Prizes are first awarded.

Pablo Picasso enters his blue period.

Frederick Soddy.

as lead, but with different atomic masses. Soddy called these substances "isotopes," and he introduced what became known as the radioactive displacement law to describe their behavior. Soddy, however, reserved the concept of isotopes for radioactive elements only. Stable elements in the periodic table, he believed, were monoisotopic.

In 1919 Aston built a new mass spectrograph that turned out to be decidedly superior to Thomson's parabola instrument. Wien and Thomson had used parallel magnetic and electric fields simultaneously to separate ions with the same value of *e/m*. The velocity of each ion determined where it fell on the parabola. Since the velocities varied, ions were spread out along the parabola rather than being focused at a specific point. Aston improved the mass spectrograph by collimating the ion beam from a discharge tube through two slits followed by mass analysis in tandem electric and magnetic fields. With this improvement ions of the same *e/m* were focused to a line on a photographic plate, and the relative abundance of the ions was indicated by the line's intensity or optical density. Ions having *e/m* ratios differing by as little as 1 part in 130 could be separated: that is, the instrument had a mass resolving power of 130. This value exceeded the resolving power of Thomson's instrument by a factor of ten.

When Aston introduced neon into his new mass spectrograph, two distinct lines appeared, one at mass 20 and the other at mass 22, along with a faint third line at mass 21. Further study led Aston to conclude that neon is composed mainly of two isotopes: one at mass 20 with an abundance of about 90 percent and the other at mass 22 with an abundance of about 10 percent. A major breakthrough was at hand. Aston's identification of the isotopes of neon extended Soddy's isotope concept to the stable elements. It also dealt a lethal blow to one tenet of Dalton's long-standing atomic theory, which held that all atoms of an element must have the same mass. In Great Britain and the United States for the next fifteen years, the most widespread application of mass spectrometry was found in the new field of isotopic analysis of the stable elements.

Aston published the results of his initial work on the isotopes of neon in November 1919. In the weeks and months that followed, he isolated isotopic species for the elements sulfur, lithium, chlorine, and silicon. By 1924 isotopes associated with fifty-three of the eighty known stable elements had been detected and their masses and abundances measured. Aston wrote prophetically that "it seems reasonable to hope that in a comparatively moderate time every stable atomic species existent in any considerable quantity on the earth will have been identified and weighed."

Of the fifty-plus elements he studied, Aston found that his measurements supported Prout's whole number rule. In 1815 William Prout had asserted that all atomic weights are whole number multiples of the weight of hydrogen. Furthermore, he assigned hydrogen special status as the substance from which all other elements are derived. This simplified view of matter was no doubt appealing to chemists and physicists, and it fit nicely with recent work on isotopes, that is, until Aston completed a second model of his mass spectrograph in 1925 and a third in 1937.

MASS SPECTROMETRY

HISTORY

1902
Léon-Phillipe Teisserenc de Bort first describes the tropospheric and stratospheric layers of the atmosphere.

Willis Carrier develops air conditioning.

1903
Lee DeForest introduces the triode ("audion")—the prototype for the modern vacuum-tube amplifier.

Aston was able to achieve a mass resolving power of 2,000 with his 1937 instrument, which far surpassed the mass resolving power of the instrument he used in 1919 to discover the isotopes of neon. More important, Aston showed that the masses of the elements and their isotopes are, contrary to his earlier observations, not *exactly* whole number multiples of the mass of hydrogen. Atomic masses up to the element fluorine, for example, are somewhat higher than the nearest whole number, while those above fluorine fall slightly below the closest integer value. At first glance these observations seemed to disprove Prout's whole number rule. However, for Aston, they further suggested that any "missing mass" might be explained by Einstein's theory of mass-energy equivalence, $E = mc^2$. The deviations from whole numbers represent the energy required to bind the atomic nucleus together. The nucleus of the helium atom, for example, consists of two protons and two neutrons. However, the mass of the helium nucleus is slightly less than the sum of the masses of two protons and two neutrons. Aston reasoned that this difference in mass in the formation of helium from its constituent subatomic particles represents what he called the "packing fraction," a measure of the stability of the atom. After plotting the packing fractions of different elements against their atomic masses, Aston generated a curve that provided important data on nuclear abundance and stability.

Aston's research on isotopes was followed closely in the United States by Arthur Dempster, a physicist at the University of Chicago. A skilled instrument builder, Dempster

Who Discovered the Isotopes of Neon?

Discoveries in science and nature might appear to be a clear-cut business. However, this is not always the case. A particularly interesting example is the case of the discovery of the isotopes of neon.

Some would argue that J. J. Thomson should be given credit for the discovery, since he saw two lines in the mass spectrum of neon. Interestingly, Thomson did not accept the idea that the less intense higher-mass line at 22 was from an isotope of neon. Instead, he proposed several other possibilities for its source. Although it had been demonstrated that radioactive elements had isotopes, he felt that the stable elements should be monoisotopic. He held firm in his belief even until March 1921 when he argued his position regarding the two lines observed in the spectrum of chlorine before the Royal Society.

Others would argue that Aston should be credited with the discovery of neon. He accepted the idea that stable elements could have isotopes. Thus, when he observed the higher mass line in the spectrum of neon in 1919, he concluded that it must be an isotope of the element neon.

Thomson saw the line at 22 in the spectrum of neon earlier than Aston did, but Aston concluded correctly the source of the line. So who discovered the isotopes of neon? You be the judge.

Reference
J. J. Thomson, et al. "Isotopes." *Proceedings of the Royal Society of London. Series A, Containing Papers of a Mathematical and Physical Character* 99:697 (2 May 1921), 87–104.

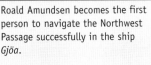
Roald Amundsen becomes the first person to navigate the Northwest Passage successfully in the ship *Gjöa*.

The Boston Red Stockings overcome the Pittsburgh Pirates in baseball's first World Series.

Orville and Wilbur Wright make the first heavier-than-air flight at Kitty Hawk, North Carolina.

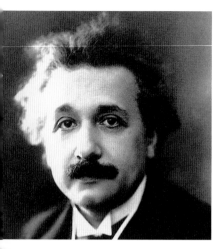

Top: Albert Einstein, shown here in his younger days.

Bottom: Arthur Dempster.

introduced a number of innovations that further advanced the study of atoms and their isotopes. In 1918 he published a paper in the *Physical Review* that described a novel method for separating ions in a magnetic field. After accelerating ions through a narrow slit, they were deflected through 180 degrees in a magnetic field, whereupon ions of a given *e/m* ratio were focused onto a narrow slit. Those ions passing through the slit were detected electrically with a quadrant electrometer in a technique similar to that employed by Thomson in his first mass spectrometer. Using his new mass spectrometer, Dempster announced the discovery of the three isotopes of magnesium in December 1920.

In the years that followed, Dempster, Aston, and other scientists working on both sides of the Atlantic Ocean completed isotopic analyses of nearly every element in the periodic table. In 1935 Dempster recorded the spectra of the last four known elements: platinum, palladium, gold, and iridium. To obtain the mass spectra of these elements, Dempster used one of his most important inventions. He had developed a new type of ion source in which a high-frequency spark was discharged between two electrodes made of the sample material to be analyzed. This device was the precursor to the spark source for ionization of solid metals.

Correlation of Mass Spectra with Molecular Structure

In the 1920s physicist John Tate, a respected authority in the field of electron ionization phenomena, began building a major research program in mass spectrometry at the University of Minnesota. At the time physicists were especially attracted to studies of molecular structure. Mass spectrometers of the type pioneered by Arthur Dempster were not well suited to study molecular structure. In 1929 Walker Bleakney—an electrical engineer who had recently joined the Minnesota physics department as one of Tate's graduate students—proposed a novel solution to this problem.

Bleakney introduced an ion source design that became a standard in mass spectrometry research. In Dempster's early mass spectrometers an electric field was used to accelerate the ions before they passed through the first slit to be separated in the mass analyzer. That same electric field was used to accelerate the beam of ionizing electrons in the ion source. Bleakney separated the fields controlling the electron and ion beams, thereby significantly improving measurements of molecular ionization and dissociation. This enhanced precision became increasingly valuable many years later as structural analysis of complex molecules, such as hydrocarbons and biological compounds, moved into the forefront of research in mass spectrometry.

When Bleakney completed his graduate studies at Minnesota in 1930, he accepted a postdoctoral fellowship from the National Research Council to pursue his investigations in mass spectrometry at Princeton University. Bleakney published more than thirty papers on mass spectrometry research during the 1930s, much of it related to his pioneering work on the isotopes of hydrogen. He confirmed the existence of deuterium, or heavy hydrogen, which had been discovered in 1932 by chemist Harold Urey at Columbia University.

The New York City subway opens.

The Russo-Japanese war breaks out.

The Herero and Hottentot peoples of Namibia revolt against German colonial rule, which ultimately leads to genocidal reprisals.

Physics department faculty, University of Minnesota, 1929. Walker Bleakney is standing in the last row, first from the left. John Tate, Bleakney's graduate adviser, is seated second from the left in the middle row.

Bleakney also obtained some of the first reliable evidence for the existence of a third isotope of hydrogen, known as tritium, which was thought to be unstable.

Further Studies of the Isotopes of the Stable Elements

Alfred Nier entered the University of Minnesota and joined Tate's group in the physics department in 1934. Nier had a master's degree in electrical engineering, which no doubt contributed to his remarkable success as a skilled craftsman of precision mass spectrometers. "When I began my career as a physics thesis student," Nier recalled years later, "a more propitious time to begin working in mass spectrometry could hardly be visualized. . . . A tradition of ingenuity in the construction of apparatus had been established in our laboratory, and high vacuum techniques were firmly established." Especially important, Nier noted, was the introduction of new electrometer vacuum tubes built by General Electric and Western Electric to measure the minuscule electric currents generated by ions striking the detector in a mass spectrometer.

1905
P. Langevin derives expressions for maximal collisional cross-sections between ions and molecules.

1905
Ernest Rutherford and Frederick Soddy introduce their theory of radioactivity.

Albert Einstein introduces his special theory of relativity.

The massacre of peaceful protesters in Moscow sparks a nationwide wave of strikes and riots.

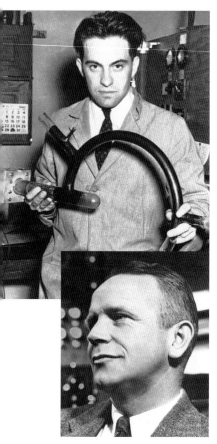

Top: A young Alfred Nier holding a mass spectrometer flight tube.

Bottom: Kenneth Bainbridge.

Nier's first mass spectrometer, which incorporated features from instruments designed by Bleakney and Tate, was used to study isotopes. Nier turned his attention to problems associated with isotopic abundance, and important results were immediately forthcoming. He confirmed the existence of ^{38}Ar and determined the abundance ratios of ^{38}Ar/^{36}Ar and ^{36}Ar/^{40}Ar. Of particular importance was Nier's discovery of the rare isotope ^{40}K, now widely used in geochronology research. In 1935 he determined the ratio of ^{40}K/^{39}K. Identification of the isotopic compositions of rubidium, zinc, and cadmium were confirmed the following year.

After he received his Ph.D. in 1936, Nier spent two years at Harvard University on a postdoctoral fellowship from the National Research Council. His collaborator during this period was Kenneth Bainbridge, whose interest in mass spectrometry began at Princeton in the late 1920s. As a postdoctoral fellow at the Franklin Institute's Bartol Research Foundation, Bainbridge had focused his efforts on the determination of precise nuclear mass measurements to confirm the theory of mass-energy equivalence proposed in 1905 by Albert Einstein.

At Harvard, Nier set to work building a new and improved mass spectrometer with a higher resolving power than the instrument he had assembled as a graduate student at Minnesota. His goal was to use this mass spectrometer to measure isotopic abundances of uranium and lead. He also carried out some of the first carbon isotope measurements, the results of which showed that the ^{13}C/^{12}C abundance ratio varies in nature. Accompanying these discoveries were major improvements in Nier's apparatus. He placed special attention on increasing the radius of curvature in the mass analyzer and improving the quality of the magnetic field. These enhancements, among others, allowed Nier to refine isotopic measurements of mercury, xenon, krypton, beryllium, iodine, arsenic, and cesium.

When he returned to the University of Minnesota in 1938, Nier once again concentrated his efforts on the study of isotopes, particularly those of carbon. The isotope ^{13}C showed potential as a tracer in studies of bacterial metabolism. Nier and his colleague, future Nobel laureate John Bardeen, built a thermal diffusion column for methane gas to enrich ^{13}C for these studies. Analysis of the samples from the bacterial metabolism studies to determine the relative abundances of ^{13}C was also carried out by Nier and his mass spectrometry group at Minnesota. "There was a great demand by our biological colleagues," he later observed, "and it quickly became apparent that more mass spectrometer capacity would be required to perform the isotopic analyses needed." However, the mass spectrometer Nier had built in Bainbridge's laboratory at Harvard used a huge two-ton electromagnet and a large five-kilowatt direct-current generator in order to produce the magnetic field in the mass spectrometer. Duplicating these components on a large scale would have been prohibitively expensive, and Minnesota did not have the financial resources that Harvard had. This situation led Nier to develop a compact, efficient, and much cheaper design that not only met the increased analytical demands of biologists but also ushered in a new era of applications in industry and academia.

MASS SPECTROMETRY	1906
	J. J. Thomson receives the Nobel Prize in physics for his investigations on the conduction of electricity by gases.

HISTORY	1905	1906	
	Norway declares its independence from Sweden.	While living in South Africa, Mohandas Gandhi leads a non-violent campaign against colonial racial policies.	The Pure Food and Drugs Act creates the U.S. Food and Drug Administration.

Origins

Alfred Nier is shown here working on his 60-degree sector field mass spectrometer, ca. 1940.

In 1940 Nier introduced his sector field mass spectrometer. This instrument marked a significant improvement over earlier devices. Rather than deflecting the ion beam through a 180-degree magnetic field, Nier's instrument incorporated a 60-degree magnet. Nier recalled that this design adjustment "greatly reduced the weight of the electromagnet and power consumption, and made possible the expanded employment of mass spectrometry for isotopic and gas analysis."

Mass Spectrometry and the Manhattan Project

In 1931 Francis Aston reported a major scientific discovery in the prestigious British journal *Nature*. Separation of ions of uranium hexafluoride gas (UF_6) in his mass spectrograph had yielded a spectral line corresponding to a mass of uranium 238. Further research along these lines led Aston to conclude that ^{238}U was the primary isotope in natural uranium ore. Four years later Arthur Dempster identified another uranium isotope, this one corresponding to mass 235. Further research by Alfred Nier at the University of Minnesota in 1939 led to the first recording of yet a third and far more rare isotope—uranium 234. Nier also determined the ratios of these three isotopes—139:1 for $^{238}U/^{235}U$ and 17,000:1 for $^{238}U/^{234}U$.

U.S. forces occupy Cuba.

San Francisco is devastated by a massive earthquake.

A letter written by Nier and his colleagues at Columbia University verifying that ^{235}U is the fissionable uranium isotope. The letter appeared in the Physical Review *in March 1940.*

A year before Nier's work on the isotopic ratios of natural uranium, Otto Hahn and Fritz Strassman, working at the Kaiser Wilhelm Institute for Chemistry in Berlin, had obtained experimental evidence of uranium fission, the process whereby uranium nuclei split apart to liberate enormous quantities of energy. At the time, however, it was unclear which uranium isotope was responsible for this process. Some physicists, led by Nobel laureate Niels Bohr, believed that ^{235}U was easily split when bombarded by neutrons. The heavier isotope, ^{238}U, on the other hand, appeared to be far more stable, often absorbing neutrons in the process of transmuting into a heavier nuclide. As the international political climate destabilized in the face of another world war, solution of what started out as a purely scientific problem soon took on a more desperate urgency. Physicists in America and Germany speculated that if the radioactive isotope in uranium fission could be identified and separated from the natural ore in sufficient quantity and purity, it might be possible to develop a weapon more powerful and destructive than any other existing at that time.

In April 1939 Nier attended the Washington, D.C., meeting of the American Physical Society, where he met Nobel laureate Enrico Fermi and an old acquaintance, Columbia University physicist John Dunning. Fermi and Dunning prompted Nier to use his 180-degree mass spectrometer at Minnesota to separate and collect the isotopes of natural uranium. With enough pure material it would be possible to determine which isotope fragmented under neutron bombardment. However, other research projects and the usual heavy teaching load of a junior faculty member prevented Nier from working on this problem until the beginning of 1940. Initial experimentation with uranium hexafluoride proved unsuccessful, largely because of its tendency to coat the exposed surfaces of Nier's mass spectrometer. After switching to uranium tetrabromide (UBr_4), "which was volatile only when heated," Nier obtained nanogram quantities of pure ^{235}U and ^{238}U. He recalled later that "this was enough so that when my Columbia colleagues . . . bombarded the targets with slow neutrons it was unambiguously clear that it was the ^{235}U that gave the fission fragments." Armed with this knowledge and especially fearful that the Nazis would exploit it first, the American government placed the fledgling civilian atomic energy program under the direct supervision of the military in 1942.

MASS SPECTROMETRY	1907		1908
	J. J. Thomson builds one of his first mass spectrographs.		Frederick Soddy investigates electrical discharges in monatomic gases.

HISTORY	1907		1908
	Fritz Haber and Carl Bosch develop their process for synthesizing ammonia.	Louis Lumière develops an early form of color photography.	A mysterious explosion occurs over Siberia's Tunguska river valley, probably the result of a meteor or comet.

Known among its thousands of participants as the Manhattan Project, this massive and secret wartime program comprised several large organizational units. Major efforts were under way at various isolated locations to develop methods that could yield large quantities of enriched ^{235}U. Scaling up separation of the two isotopes, however, was an entirely different matter that would prove exceedingly difficult. Especially problematic was the fact that ^{235}U and ^{238}U differed in mass by a little more than one percent, which significantly increased the complexity of the various separation techniques under consideration: electromagnetic separation, gaseous diffusion, and high-speed centrifugation.

By the summer of 1943 enrichment by preparative scale mass spectrometry was by far the most promising of the three uranium purification techniques being considered. Two years earlier Ernest Lawrence, a physicist who directed the electromagnetic separation program at Berkeley, modified his cyclotron so that it operated as a 180-degree Dempster-type mass spectrometer. The source used an electron beam to ionize the vapor of a uranium salt.

In February 1942 Lawrence made several major improvements to the design of his modified cyclotron. He replaced the cyclotron tank with a new evacuated chamber shaped like the letter *C* to match the semicircular path of the ion beam as it traversed the magnetic field. The ion source was located at one end of the chamber, and the collector was placed at the opposite end. One immediate advantage of the new design was its reduced size, which allowed for the placement of more than one unit between the magnet faces. Within a short

Ernest Lawrence makes adjustments to his thirty-seven-inch cyclotron at the University of California at Berkeley.

1909

Robert Millikan and Harvey Fletcher determine the charge of the electron.

1909

The "Young Turks" win democratic reform and constitutional government in the Ottoman Empire.

The Ford Model T is introduced.

The National Association for the Advancement of Colored People (NAACP) is founded in New York City.

time Lawrence's first generation "calutron" (short for *Cal*ifornia *U*niversity cyclo*tron*) had produced one hundred micrograms of pure ^{235}U, which was sufficient to determine the properties of the material and demonstrate that electromagnetic separation was a suitable production technique.

While Lawrence was still collecting micrograms of ^{235}U in his thirty-seven-inch cyclotron, workers in eastern Tennessee were busy constructing the cavernous buildings that would contain the hundreds of calutron tanks needed to produce large quantities of purified uranium metal. Installation of the first "alpha racetrack" at Oak Ridge, consisting of ninety-six vertically oriented calutrons arranged in the shape of an oval, was completed by the end of 1943. Early in 1945 hundreds of calutrons operated simultaneously to separate on a daily basis just over seven ounces (198.8 grams) of 80-percent-enriched ^{235}U, which was of sufficient purity to make an atomic bomb.

Like electromagnetic separation, gaseous diffusion was a simple concept on paper, but it turned out to be an engineering nightmare. The idea, pioneered by chemist Harold Urey and his research staff at Columbia University in 1941, held that if uranium hexafluoride gas were pumped through a porous barrier, ^{235}U would pass through more easily than the heavier ^{238}U. However, it was necessary to push the gas through a number of barriers to achieve sufficient separation of the two isotopes. The final design used at Oak Ridge called for thousands of barriers, each fashioned in the shape of an airtight tank and connected to one another. Construction of K-25, the designation assigned to the mammoth gaseous diffusion plant at Oak Ridge, began in the spring of 1943.

Calutron "racetracks" like this one at Oak Ridge produced enriched ^{235}U for the world's first atomic weapons.

Because of the corrosive effects of uranium hexafluoride, engineers assembling the pumps and diffusion barriers were confronted by strict requirements to prevent leaks into the process stream from the atmosphere. UF_6 was highly sensitive to contaminants, even minute quantities of water vapor in air. The presence of trace amounts of organic materials was equally problematic. The vacuum pump oils, for example, were vulnerable to attack by UF_6. One small leak anywhere along the process line, and the diffusion barriers would gum up and fail, shutting down the entire system. Nier himself observed that "conventional leak testing methods were not sensitive enough to meet the stringent requirements we faced." Of the various methods available at the time, two were especially

MASS SPECTROMETRY	1910	
	J. J. Thomson detects secondary emission.	Cambridge researcher H. Thirkill investigates so called magneto cathodic rays.

HISTORY	1909	1910	
	Robert Edwin Peary leads the first expedition to the North Pole.	The Union of South Africa, formed from several former British colonies, becomes an independent state.	Japan formally annexes Korea.

The K-25 gaseous diffusion plant, Oak Ridge, Tennessee.

common: detecting small pressure changes when a system suspected of leaking was sealed with wax and monitoring changes in the apparent pressure in a system sprayed with a liquid (ether, CCl_4) or a gas (CO_2, H_2, He). Nier adapted his sector mass spectrometer to make a highly sensitive helium leak detector that General Electric then mass-produced for use in the gaseous diffusion plant. Nier's instrument used a vastly superior method of leak detection and was well suited to the stringent requirements for leak-tight cells in the gaseous diffusion plant.

In addition to the helium leak detectors manufactured by GE, Nier-type sector mass spectrometers were also used in the gaseous diffusion plant to monitor the uranium process streams. These instruments measured and recorded on a continuous basis the principal constituents of the process stream. Closely monitored by one hundred mass spectrometers—fifty on-line and fifty as backup instruments—the K-25 plant went into operation in January 1945. The world's first uranium weapon, equivalent to the explosive force of nearly nineteen thousand tons of TNT (trinitrotoluene), was detonated in the New Mexico desert just six months later.

The major developments in mass spectrometry during the Manhattan Project were published in the scientific literature after the war. This work made it clear that mass spectrometry was crucial to the success of the project and that it was a useful tool in four different areas: as a general-purpose analytical tool capable of quantifying components of a mixture; as a

1911

Wilhelm Wien receives the Nobel Prize in physics for his discoveries concerning the laws of heat radiation.

1911

Revolution breaks out in Mexico. It will last ten years and leave one million dead.

Wassily Kandinsky paints *First Abstract Watercolor*, considered the first truly abstract painting.

Victor Hess discovers cosmic rays by using balloons.

A General Electric leak detector.

preparative scale tool for the separation of isotopes; as a process control tool capable of monitoring a process stream; and as a tool for the detection of extremely small leaks in vacuum chambers. The demonstrated prowess of the mass spectrometer in each of these four areas found application in the postwar era, although the general-purpose analytical application and leak detection in vacuum chambers are the most commonly known applications today.

Suggested Reading

F. W. Aston. "Constitution of Thallium and Uranium." *Nature* 128 (1931), 725.

———. *Isotopes.* London: Edward Arnold, 1924.

———. *Mass Spectra and Isotopes.* London: Edward Arnold, 1942.

———. "A New Mass Spectrograph and the Whole-Number Rule." *Proceedings of the Royal Society of London, Part A* A115 (1927), 487–514.

———. "A Positive Ray Spectrograph." *Philosophical Magazine* 38 (1919), 707–715.

W. H. Brock. *The Norton History of Chemistry.* New York: W. W. Norton, 1992.

P. F. Dahl. *Flash of the Cathode Rays: A History of J. J. Thomson's Electron.* Bristol, England: Institute of Physics Publishing, 1997.

E. A Davis; I. J. Falconer. *J. J. Thomson and the Discovery of the Electron.* London: Taylor & Francis, 1997.

A. J. Dempster. "Isotopic Constitution of Uranium." *Nature* 136 (1935), 180.

———. "A New Method of Positive Ray Analysis." *Physical Review* 11 (1918), 316–324.

R. G. Hewlett; O. E. Anderson, Jr. *The New World, 1939–1946.* Vol. 1 of *A History of the United States Atomic Energy Commission.* University Park: Pennsylvania State University Press, 1962.

N. W. Hunter. "Francis William Aston." In *Nobel Laureates in Chemistry,* edited by L. K. James, 140–145. Washington, D.C.: American Chemical Society; Philadelphia: Chemical Heritage Foundation, 1993.

N. W. Hunter; R. Roach. "Frederick Soddy." In *Nobel Laureates in Chemistry,* edited by L. K. James, 134–139. Washington, D.C.: American Chemical Society; Philadelphia: Chemical Heritage Foundation, 1993.

A. J. Ihde. *The Development of Modern Chemistry.* New York: Dover Publications, 1984.

A. O. Nier. "The Isotopic Constitution of Uranium and the Half-Lives of the Uranium Isotopes." *Physical Review* 55 (1939), 150–153.

———. "Some Reflections on the Early Days of Mass Spectrometry at the University of Minnesota." *International Journal of Mass Spectrometry and Ion Processes* 100 (1990), 1–13.

———. "Some Reminiscences of Isotopes, Geochronology, and Mass Spectrometry." *Annual Review of Earth and Planetary Sciences* 9 (1981), 1–17.

———. "Some Reminiscences of Mass Spectrometry and the Manhattan Project." *Journal of Chemical Education* 66 (1989), 385–388.

A. O. Nier; J. Bardeen. "The Production of Concentrated Carbon (13) by Thermal Diffusion." *Journal of Chemical Physics* 9 (1941), 690–692.

A. O. Nier et al. "Mass Spectrometer for Leak Detection." *Journal of Applied Physics* 18 (1947), 30–33.

———. "Nuclear Fission of Separated Uranium Isotopes." *Physical Review* 57 (1940), 546.

MASS SPECTROMETRY

HISTORY | **1911**

The ancient Inca city of Machu Picchu is discovered by Hiram Bingham.

Roald Amundsen of Norway leads the first expedition to reach the South Pole.

John D. Rockefeller's Standard Oil Trust is broken up in an antitrust suit.

————. "Recording Mass Spectrometer for Process Analysis." *Analytical Chemistry* 20 (1948), 188–192.

J. R. Partington. *A History of Chemistry.* Vol. 4. London: MacMillan, 1964.

R. V. Pound; N. F. Ramsay. "Kenneth Tomkins Bainbridge." In *Biographical Memoirs of the National Academy of Sciences, 1877–*, Vol. 76, 18–34. Washington, D.C.: National Academy Press, 1999.

G. T. Reynolds. "Walker Bleakney." In *Biographical Memoirs of the National Academy of Sciences, 1877–.* Vol. 73, 86–99. Washington, D.C.: National Academy Press, 1998.

J. H. Reynolds. "Alfred Otto Carl Nier." In *Biographical Memoirs of the National Academy of Sciences, 1877–*, Vol. 74, 244–265. Washington, D.C.: National Academy Press, 1998.

R. Rhodes. *The Making of the Atomic Bomb.* New York: Simon & Schuster, 1986.

A. J. B. Robertson. *Mass Spectrometry.* London: Methuen, 1954.

H. D. Smyth. *Atomic Energy for Military Purposes: The Official Report on the Development of the Atomic Bomb under the Auspices of the United States Government, 1940–1945.* Princeton, N.J.: Princeton University Press, 1945.

H. J. Svec. "Mass Spectrometry—Ways and Means: A Historical Prospectus." *International Journal of Mass Spectrometry and Ion Processes* 66 (1985), 3–29.

A. L. Yergey; A. K. Yergey. "Preparative Scale Mass Spectrometry: A Brief History of the Calutron." *Journal of the American Society for Mass Spectrometry* 8 (1997), 943–953.

Direct Quotations

Page 8, Aston, 1924, p. 7.

Page 11, Nier, 1990, pp. 6–7.

Page 12, Nier, 1990, p. 11.

Page 13, Nier, 1990, p. 12.

Page 14, Nier, 1989, pp. 386–387.

Page 16, Nier, 1989, p. 388.

1912

Richard Whiddington investigates the transmission of cathode rays through matter.

1912

The *Titanic* sinks in the North Atlantic after striking an iceberg.

Alfred Wegener proposes the idea of continental drift.

U.S. forces occupy Nicaragua and Honduras.

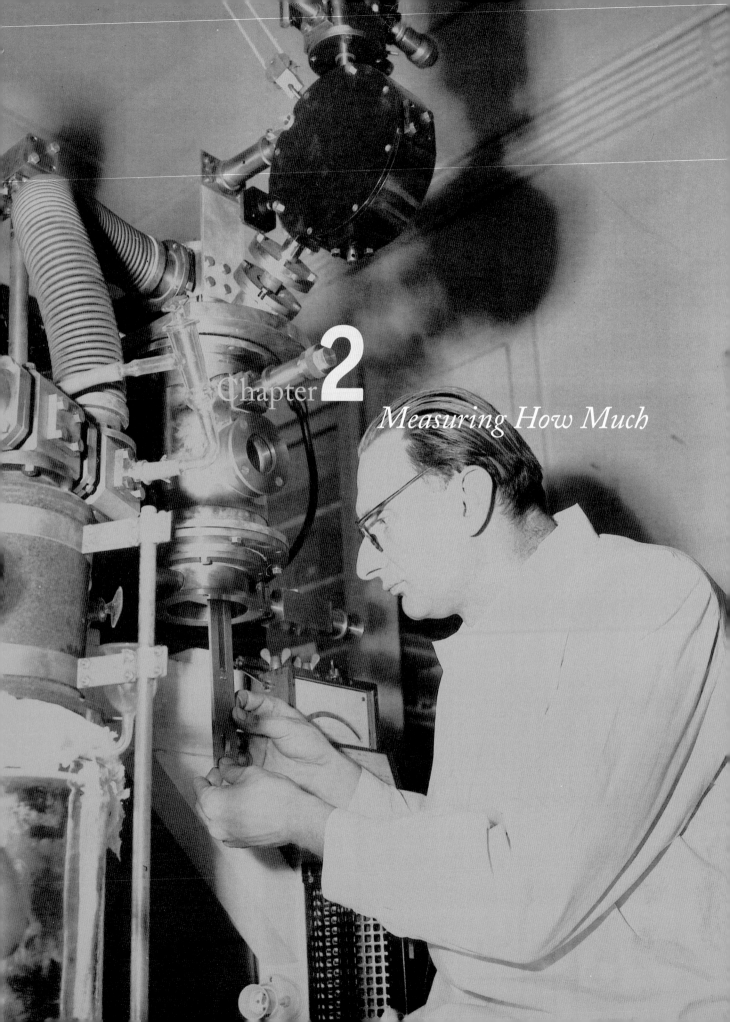

Chapter **2**

Measuring How Much

Josef Mattauch (right) and Richard Herzog (left) pioneered the development of the double-focusing mass spectrograph. Herzog is shown here handling a vacuum spectrograph at Vienna University's Institute of Physics, 1952.

After World War II, physicists continued to exploit the analytical capabilities of mass spectrometry to measure precisely the exact masses of the elements and relative abundances of the isotopes. This important work was carried out to a large extent in a handful of academic institutions, including Harvard University, the University of Minnesota, and the University of Vienna.

Interest in precise mass measurement was driven at least in part by important improvements in instrumentation dating back to the 1930s. During this period precision mass analysis was carried out using two types of instruments—the mass spectrometer and the mass spectrograph. Although the mass spectrograph enjoyed widespread use in this field of research before and immediately after World War II, rapid advances in electronics technology after the war helped the mass spectrometer supplant the mass spectrograph as the instrument of choice among researchers interested in accurate mass determinations.

Much of the improvement in mass spectrometers for precision mass analysis can be attributed to Alfred Nier. His contributions improved the resolution and sensitivity of mass spectrometers to a level of performance comparable to that of the most advanced mass spectrographs built by his Harvard mentor Kenneth Bainbridge and the Austrian team of Josef Mattauch and Richard Herzog. After World War II, Nier set to work with his graduate student Edgar Johnson to build a second-order double-focusing mass spectrometer. Introduced in 1953, this advanced instrument used a 90-degree electrostatic analyzer followed by a 60-degree magnetic analyzer to achieve high sensitivity and improved

MASS SPECTROMETRY	1913		
	J. J. Thomson observes a line at mass 22 in the spectrum of neon.	J J. Thomson shows that reaction of H_2^+ with a hydrogen molecule occurs via proton transfer.	J. J. Thomson delivers his Bakerian Lecture, "Rays of Positive Electricity," to the Royal Society of London.

HISTORY	1913		
	Albania gains independence from the Ottoman Empire.	Igor Stravinsky's and Vaslav Nijinsky's ballet *The Rite of Spring* is first performed in Paris.	Earth's ozone layer is discovered by Charles Fabry of France.

All Mass Spectrometers Are Divided into Five Parts

Regardless of the type of instrument or the type of analysis, a mass spectrometer is made up of five separate systems: inlet, ion source, mass analyzer, detector, and recorder.

Since mass spectrometers operate in a vacuum, an inlet system is required to transfer the sample from ambient room pressure into the ion source. The ion source converts the sample molecules into sample

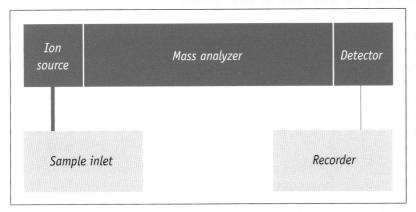

ions. An ion is a molecule with an electrical charge. The simplest sample ion is formed by removing an electron from the neutral molecule, thus forming a positive ion. Many different kinds of ion sources have been developed over the years, but the important

point is that sample molecules must be made into sample ions before mass analysis can be performed.

In the mass analyzer, magnetic or electric fields, or both, are used to control the movement of sample ions so that they are separated from each other on the basis of their mass. There are many different kinds of mass analyzers, but they all perform the same function: to separate ions according to their mass.

The detector senses the arrival of mass-separated ions and amplifies the minuscule ion current for further electronic processing. There are many different kinds of detectors, but they all have the task of detecting ions and converting them to a stronger electronic signal.

The recorder accepts the signal from the detector, amplifies it further, and records it, thus creating the mass spectrum. The recorder can be as simple as a strip chart recorder, or it can be as complex as a high-powered desktop computer. In either case the data from the recorder can be stored indefinitely and recalled as necessary for further analysis.

As noted above, a mass spectrometer operates in a vacuum. Generally the vacuum is in the range of several millionths to one billionth of the ambient pressure in the room. The ion source, mass analyzer, and detector are under vacuum, as denoted by the blue color in the diagram on the left.

resolving power. These features improved mass measurement accuracy. A variation of the Nier-Johnson geometry for precision mass analysis was exploited in a number of very successful commercial instruments designed and built by Associated Electrical Industries (now Kratos Analytical Instruments) and marketed in various versions from the early 1960s through the 1990s.

Isotopes as Clocks

While physicists were using mass spectrometry to determine and refine the masses of the elements and their isotopes, scientists in other fields recognized the potential of mass spectrometry to provide useful information. Beginning in the late 1930s, interactions among

HISTORY

1913

Henry Ford introduces the assembly line to automobile manufacturing.

1914

World War I begins.

scientists led to the use of mass spectrometry in two important but separate disciplines, geology and biology. The geological connection began when Alfred Nier was at Harvard working with Kenneth Bainbridge on the isotopes of lead. There were actually two separate biological connections, one after 1932 when Harold Urey provided his Columbia University colleagues Rudolf Schoenheimer and David Rittenberg with deuterium as a tracer in their metabolic studies, and the other in 1940 when Nier provided biochemist Harland Wood with some ^{13}C for his studies of bacteria at Iowa State University. The use of isotopically enriched compounds to explore biological processes will be discussed in greater detail in chapter 5. Here we will explore the application of mass spectrometry in the field of geology.

Before Nier joined Bainbridge's laboratory in 1936, it had been known for some time that the decay of uranium and thorium to lead in an object of geological interest could be used as an internal clock to determine its age. At the time wet chemical methods were the only means of determining the atomic weight of an element. These methods actually measured the chemical weight of the element and gave limited information about the isotopic distribution. Using the 180-degree mass spectrometer he built with Bainbridge's assistance, Nier examined the processes by which natural uranium decayed into lead. Measurement of the lead content in a geological specimen and knowledge of the rate of decay of uranium to lead would, after running a simple calculation, yield the age of the mineral sample.

As Nier soon realized, however, the analytical process would not be so simple. The specimen always contains a certain amount of common lead not associated with radioactive decay. Thus, simple measurement of the amount of lead in the sample would not yield an accurate result. Here the mass spectrometer played a crucial role. The two uranium isotopes, ^{238}U and ^{235}U, decay into ^{206}Pb and ^{207}Pb, respectively. Common lead, meanwhile, contains these two isotopes as well as two additional isotopes—^{204}Pb and ^{208}Pb—in relatively constant proportions. If the isotopic composition of the common lead could be determined and subtracted from the amount of ^{206}Pb and ^{207}Pb present as a result of the radioactive decay of uranium, it would be possible to calculate a more accurate value for the age of the specimen. The same technique could be used for other radioactive elements as well, including thorium.

Mass spectrometry was superior to any existing wet chemical analytical method used to determine the mass of an element in a sample. It was also faster and required a much smaller amount of sample material. This work was instrumental in providing geologists with a powerful tool that has since been used to investigate a variety of geochronological problems, related both to age dating and to investigating the history of events in our solar system and on Earth. The use of isotopic analysis by mass spectrometry has expanded to other radioactive transitions with different decay times suitable for dating objects over a variety of time frames.

While the measurement of the relative abundances of the elements and their isotopes in an object submitted for age dating are quite precise, interpreting the results is the subject of much discussion among scientists. Generally, geologists must make assumptions about the past history of an object when interpreting aging determinations by mass spectrometric data.

Top: Edgar Gustav Johnson.

Bottom: Harland Wood.

1915

German researcher T. Retschinsky studies magnetic spectrum of oxygen canal rays.

1915

Ernest Rutherford discovers the proton.

The Panama Canal is opened to sea traffic.

Franz Kafka publishes his short story "The Metamorphosis."

Growing Sector Instruments

Early sector instruments consisted of a single magnetic sector for separation of mass according to the momenta of the ions. This arrangement was satisfactory to correct for beam aberrations arising from variations in the spatial location of ions in the ion source, but the need for greater mass measurement accuracy eventually arose. In particular, the velocity distribution of ions leaving the ion source limited the resolution that could be achieved with a single magnetic sector. Consequently, instrument designers turned to the use of multiple sectors. These double-focusing instruments consisted of two separate sectors in tandem through which the ions had to pass. The most popular arrangement was one with an electric sector (E) followed by a magnetic sector (B), forming what is known as an EB instrument. By careful analysis of the equations of motion, the geometry of the two sectors could be arranged so that the mass spectrometer would be double focusing, that is, second-order aberrations of the ion beam would be corrected.

Today, geologists accept that the age of Earth, the Moon, and meteorites is the same, approximately four-and-a-half billion years. By measuring strontium isotopic ratios from the decay of rubidium in lunar objects, they have been able to determine that some lunar rocks were formed a little more than three-and-a-half billion years ago, suggesting that some portion of the Moon must have undergone a major disturbance, such as a meteorite impact or volcanic activity, within a billion years of its formation. These same techniques can be used to provide information about the dynamics of Earth, such as the growth rates of the continental plates, the flow of material from mid-oceanic ridges, and the mixing of material at subduction zones.

Isotopes as Thermometers

An example of some of the other studies that make use of precise mass analysis of isotopic abundances is found in estimates of the surface temperature of Earth over the past hundred million years. Scientists know that the relative abundance of ^{18}O to ^{16}O in ocean water varies slightly with temperature. Shellfish incorporate this abundance information in the calcium carbonate that makes up their shells. Analyzing the ratio of the two oxygen isotopes in fossilized shells from deep-sea cores permits estimation of Earth's temperature at the time the organism was alive. Temperature determinations from this type of analysis are referred to as isotope temperatures. Information from these studies sheds light not only on past glaciation events but also on the effect of plate tectonics on climate.

An international team recently studied ice cores from Antarctica to measure in detail the isotope temperature during the last four glaciation events. Combined with other information, such as dust and sodium content variations and the distribution of carbon dioxide and methane throughout the ice core, a hundred-thousand year cycle is easily discerned in the data. Of particular interest is the strong correlation between atmospheric greenhouse gas concentrations and the isotope temperature. Climatic simulations based on the ice core results suggest that greenhouse gases contribute significantly to the average glacial—

MASS SPECTROMETRY

1916

Arthur Dempster proposes that electron impact ionization of hydrogen "seems to involve only the detachment of an electron" from the hydrogen molecule.

HISTORY

1915

William Draper proposes that hydrogen might fuse into helium with the release of energy.

The first coast-to-coast telephone call is made between New York and San Francisco.

1916

Turkish forces systematically murder more than one million Armenians in one of the first modern acts of genocide.

interglacial temperature change. This observation is particularly noteworthy, since modern levels of both methane and carbon dioxide are in excess of peak values observed in the last four glaciation periods.

High-Precision Isotope Ratio Measurements

In 1939 Alfred Nier and his colleagues at the University of Minnesota showed that carbonates have a greater $^{13}C/^{12}C$ ratio than other natural carbon sources, thus ushering in the era of what is now called high-precision isotope ratio mass spectrometry (IRMS). IRMS provides a dimension of information that is radically different from the other two types of mass spectral information: structure determination and quantitative analysis. Subtle isotope effects are ubiquitous in chemical and physical processes because the rates of change for the heavier isotopes are slightly different than those for the lighter isotopes. Thus extremely precise quantitative isotopic analysis can be used to extract valuable information about a wide variety of natural processes.

Instruments used for high-precision IRMS are designed for optimal isotopic analysis of the gases CO_2, H_2, N_2, and SO_2, and in special cases of a few inert gases. In return for this inflexibility, isotope ratios of $^{13}C/^{12}C$, $^{18}O/^{16}O$, $^{2}H/^{1}H$, $^{15}N/^{14}N$, and $^{34}S/^{32}S$ can be determined with up to seven significant figures. This remarkable precision depends on the use of one of two styles of inlet systems: the classic dual inlet, introduced in 1947, or the continuous-flow inlet, introduced thirty years later.

The dual inlet was first integrated into a general-purpose isotope ratio mass spectrometer for high-precision measurements of biological processes in 1950. Using this instrument in the early 1960s, researchers at the California Institute of Technology discovered that plants use carbon isotopes through two different mechanisms, now known as the C3 and C4 photosynthetic pathways. This discovery is still used extensively to identify food sources of modern and ancient animals and humans in archaeological studies, to measure subtle changes in the $^{13}C/^{12}C$ ratio in the food chain, and to determine the interactive effects of enzymes on carbon isotope ratios in specific biological materials, such as fatty acids.

The recent development of continuous-flow inlet systems for IRMS enabled tremendous advances in precision mass analysis. Continuous-flow IRMS permits more rapid analysis than can be achieved with instruments equipped with dual-inlet systems, and it can also handle much smaller quantities of sample. The primary drawback of continuous flow is its limited precision, although it is still sufficient for most applications. Beginning in the 1970s, the continuous-flow inlet was adopted for use in gas chromatograph–mass spectrometers in order to carry out precision isotopic analysis of CO_2 and N_2. A decade later the first elemental analyzer, a device that converts solid samples into CO_2 and N_2 gas for quantitative analysis, was interfaced to a mass spectrometer for isotope ratio studies.

More recently researchers have been able to exploit these innovations in instrumentation to use IRMS for myriad applications. The introduction of the elemental analyzer, for

Jeanette Rankin of Montana becomes the first woman elected to the U.S. House of Representatives.

Margaret Sanger opens the first birth control clinic in the United States.

Albert Einstein publishes his general theory of relativity.

Standardizing Mass

Until the 1920s oxygen (atomic weight 16) served as the effective standard of comparison for all the chemical elements and their isotopes for both physicists and chemists. Specialists in the new field of mass spectrometry were especially active in using this standard to determine the exact masses of the stable elements based on oxygen having a mass of 16.00 atomic mass units. In 1929, however, W. F. Giauque and H. L. Johnson at the University of California discovered that oxygen is not monoisotopic and that two very low concentration isotopes, ^{17}O and ^{18}O, exist in the atmosphere. This subtle difference created a serious problem. Since chemists measure the atomic weight of the elements by macroscopic, wet chemical techniques, their measure of the atomic weight of oxygen included not only the most abundant isotope at ^{16}O but also the two less abundant isotopes. However, physicists were able to focus their measurements on only the most abundant isotope of oxygen, ^{16}O. As a result chemists and physicists ended up with slight differences in the masses of the elements determined on the basis of oxygen as a standard. Even though researchers in both disciplines were using the same element as the standard against which to measure the masses of the other elements, there was a difference of 0.0044 atomic mass units in the standards, which led to an unsatisfactory situation in which each group had their own tables for the masses of the elements with slightly different values. Complicating matters further, it was discovered that the isotopic composition of natural oxygen varied slightly around the world.

The international scientific community recognized the need to establish a single scale for both chemical and physical atomic weights, and in 1956 the International Union for Pure and Applied Chemistry (IUPAC) requested that inquiries be made into the possibility of unifying the O = 16 scale used by chemists and the ^{16}O = 16 scale used by physicists. Any change to either standard was complicated by the fact that it would require the other group to make adjustments to vast bodies of tabulated data, and scientists in both disciplines were understandably reluctant to undertake this monumental task. Given the fact that the isotopic composition of oxygen varied according to location, it was obvious that some other element should be chosen as the standard.

Many proposals were discussed among the physicists and chemists concerning other elements that could be used as a common standard for both communities. The most attractive alternative was proposed when physicists Josef Mattauch and Alfred Nier suggested a new standard based on ^{12}C = 12. Mattauch campaigned extensively both in Europe and the United States for the carbon-12 standard. While a standard based on carbon seems at first counterintuitive, since 1.1% of all carbon is ^{13}C, the advantages of using ^{12}C outweighed the requirement that chemists would have to use isotopically purified carbon to perform atomic weight determinations by wet chemical methods. Besides, by this time, mass spectrometry was becoming an important analytical tool in the chemistry community, and wet chemical methods of determining atomic weight of the elements were falling out of favor. Carbon-12 was already the most important secondary standard for the mass spectroscopic determination of atomic weights. The O = 16 standard required that researchers spend a good deal of their time quantifying the $^{12}C/^{16}O$ ratio to a high degree of accuracy. The ^{12}C = 12 scale would free up practitioners in mass spectrometry to spend their time on other research questions. Finally, the adjustments that needed to be made to existing tables of data were small and were borne equally by both the chemical and physical communities.

The overwhelming advantages of the ^{12}C = 12 scale led the General Assembly of IUPAC to accept it in 1959, followed by the General Assembly of the International Union of Pure and Applied Physics in 1960. This joint ratification allowed the two disciplines to share results again using a common scale for the mass of the elements. Thus, a vexing problem that had been festering for decades was resolved in a way that was advantageous to both communities.

MASS SPECTROMETRY	1917
	German physicist G. Wendt studies multiple charge formation in canal rays of carbon, silicon, and boron.

HISTORY	1916	1917	
	The Easter Uprising in Ireland is crushed by British forces.	Karl Schwarzchild predicts the existence of black holes in outer space.	Revolution brings the Communists to power in Russia.

example, helped revolutionize several fields, ranging from ecology to food analysis. In a typical example a systematic change in hydrogen and oxygen isotope ratios with latitude, known as the meteoric water line, was discovered using high-precision IRMS. This change is caused by isotope effects of evaporation and condensation between the Earth's equator and the polar caps. Identifying the natural geographic segregation of isotopes has also permitted rigorous evaluation of the migratory history of several animal species. High-precision IRMS studies of monarch butterfly wings, which are formed in the larval stage, can be used to establish the migratory origin of butterflies wintering in Mexico. In addition, this technique can be used to track migratory patterns of birds and whales. Geographic and other factors that induce isotopic differences in bodies of water permit evaluation of water source utilization in trees.

Accelerator Mass Spectrometry: More with Less

The method of radiocarbon dating that measures the radioactive decay of ^{14}C was introduced by chemist Willard Libby almost sixty years ago. This dating method for carbon was chosen over the mass spectrometric method introduced by Nier for uranium-lead because of

Willard Libby.

several problems unique to carbon. First, the amount of ^{14}C is very small and thus is difficult to quantify even with the most sensitive mass spectrometer detectors. However, an even more serious problem is the fact that ^{14}N will always be present in any mass spectrometer because of the small amount of air in the vacuum system. Measuring a minuscule amount of one component (^{14}C) in the presence of a large amount of another component (^{14}N) is unlikely to provide useful quantitative results.

This operational difficulty disappeared almost overnight when, in 1977, researchers working simultaneously at several American universities described what is now known as accelerator mass spectrometry (AMS). In this technique a high-energy accelerator is used to accelerate ions to energies as high as sixty million volts. Under these conditions, the more sensitive ion counting techniques of nuclear instrumentation can be used for detection. When this experiment is performed with positive ions, the problem of ^{14}N interference remains. However, the problem disappears when measurements are taken using negative ions, since negatively charged ^{14}N ions are not stable and do not survive to reach the detector. Using this approach, a group of academic and industrial researchers in the United States and Canada showed that it was possible to measure as few as three ^{14}C atoms in more than 10^{16} ^{12}C atoms. One practitioner noted that accelerator mass spectrometric methods are two hundred thousand times more sensitive than conventional radiocarbon methods for determining the age of artifacts containing carbon.

1918

A. J. Dempster develops a 180-degree magnetic-focusing mass spectrometer that uses electron ionization and thermal ionization sources.	German scientist W. Volker produces alkali metal ions from salts under bombardment by canal rays.

1918

The collision of two ships, one transporting military explosives, in the harbor of Halifax, Nova Scotia, results in an explosion that leaves 1,600 dead.	World War I ends.	Tokyo is hit by a massive typhoon, and 1,600 die.

Perhaps the most important and useful feature of AMS is that it only requires milligram quantities of material, thus permitting researchers to extract samples from extremely rare and valuable artifacts. Since the introduction of accelerator mass spectrometry for carbon dating, the tool has been applied to a number of other radioactive transitions with even longer half-lives, such as ^{10}Be, ^{26}Al, ^{36}Cl, ^{41}Ca, and ^{129}I. This technique provides scientists with the means of dating objects from a variety of sources over hundreds of thousands of years.

From a Handful of Elements to the Rest

As we have seen in this chapter, isotope ratio mass spectrometry focused on precision measurements of a few select elements. However, inorganic elemental mass analysis has also been used to study the much broader range of elements found in metals, alloys, and semiconductors. As usual, the difficulty of analysis by mass spectrometry centers on the development of an appropriate technique for ionizing the sample. After World War II mass spectrometrists endeavored to overcome this obstacle by developing effective ion sources for use in the mass analysis of inorganic materials.

The most commonly used ionization method for elemental analysis by mass spectrographs initially was the spark source first introduced by Arthur Dempster in 1936 to create ions from electrically conductive samples. However, it was more difficult to apply this ionization technique to semiconductors and insulating materials. While spark source ionization was used with a fair degree of success for this class of materials through the 1960s, its use declined to the point at which it is barely used today. This decline came about for several reasons, primarily because of the experimental difficulties of performing quantitative elemental analysis with the spark source and a corresponding increase in the use of other techniques of elemental analysis.

With the decline of the spark source as a means of ionizing samples for elemental analysis, interest turned to alternative ionization methods. Developments in the glow discharge ion source enabled the use of mass spectrometers, which used electrical detection rather than mass spectrographs for elemental analysis. However, the glow discharge method of ionization had an inherent disadvantage for ultra-trace analysis in that there was some carryover of the previous sample to the next sample.

Some researchers developed laser ionization mass spectrometry, which eliminated carryover between samples and their associated memory effects. However, it was found that an even better ionization method was to have the plume of material ablated from a laser shot expand into an inductively coupled plasma (ICP)—an ionized gas. In this case neutral molecules from the sample in the laser plume are ionized as well, thus enhancing the sensitivity of the technique. With the passage of time the advantages of the ICP ion source for inorganic analysis were recognized and developed for an ever-expanding range of applications.

One of the advantages of the laser as a means of ionization is that it can be focused onto a very small area of the surface of the sample, thereby obtaining spectra only from the

MASS SPECTROMETRY	1919
	German physicist H. Baerwald proposes that the radiation proceeding from incandescent tungsten has a constitution similar to that of canal rays.

HISTORY	1918–19	1919	
	A worldwide influenza epidemic kills an estimated 30 million people.	Arthur Eddington first observes the bending of light by gravity.	The League of Nations is founded.

Plasma Ionization Techniques

Plasma ionization techniques are used to ionize inorganic solids, such as metals and alloys. Plasmas can be generated at extremely low pressures—about one tenth of one percent of atmospheric pressure—using a glow discharge apparatus. At atmospheric pressures plasmas can be created by using radio-frequency energy to excite a gas.

In the case of a glow discharge ion source, about a thousand volts are placed across the gap between the negatively charged cathode and the positively charged anode. Low-pressure gas, such as argon or xenon, is then introduced into the tube. Current flow in the glow discharge is restricted to just a few milliamperes. The sample material either constitutes the cathode itself or is coated onto a cathode made of dif-

ferent material. As the current flows through the discharge between the gap, sample ions are produced.

At higher pressures the inductively coupled plasma source is often used. A load coil couples the radio-frequency energy to the gas, typically argon. The sample, usually in the form of a gas or liquid, is added to the plasma, where it is ionized. Introducing the sample ions and plasma into the mass analyzer presents some difficulties, since the plasma is at atmospheric pressure and is maintained at a temperature in the vicinity of 5,300 °C. The introduction of the ionized sample into the mass spectrometer is accomplished by a series of water-cooled skimmers separated by differentially pumped regions between the plasma source and the mass analyzer.

constituents of that small area. By moving the beam across the sample in an orderly fashion (rastering) and recording the mass spectra at each location, information about the spatial distribution of compounds on the surface can be determined. This same approach is also used in secondary ion beam mass spectrometry (SIMS), in which a beam of ions—such as argon, oxygen, cesium, or gallium—is focused on the surface to ablate it and create secondary ions of the surface material. In some cases a conventional electron ionization beam is used to ionize any neutral molecules in the plume of material created in SIMS, once again enhancing the technique's sensitivity.

The application of mass spectrometry to precision isotopic and inorganic elemental analysis has been extraordinarily productive since the first experiments were done more than half a century ago. These wonderful instruments promise even greater discoveries in the future.

Suggested Reading

F. Albarede. "Sm/Nd Constraints on the Growth Rate of Continental Crust." *Techtanophysics* 161 (1989), 299–305.

F. W. Aston. *Mass Spectra and Isotopes.* London: Edward Arnold, 1942.

C. L. Bennett et al. "Radiocarbon Dating Using Electrostatic Accelerators: Negative Ions Provide the Key." *Science* 198 (1977), 508–510.

P. B. Best; D. M. Schell. "Stable Isotopes in Southern Right Whale (*Eubalaena australis*) Baleen as Indicators of Seasonal Movements, Feeding and Growth." *Marine Biology* 124 (1996), 483–494.

W. A. Brand. "High Precision Isotope Ratio Monitoring Techniques in Mass Spectrometry." *Journal of Mass Spectrometry* 31 (1996), 225–235.

1920

A German group studies the effect of cathode rays on bacteria and larva.

Afghanistan officially gains independence from Britain.

1920

Women in the United States gain the right to vote with the ratification of the 19th Amendment to the Constitution.

J. T. Brenna et al. "High-Resolution Continuous-Flow Isotope Ratio Mass Spectrometry." *Mass Spectrometry Reviews* 16 (1997), 227–258.

C. P. Chamberlain et al. "The Use of Isotope Tracers for Identifying Populations of Migratory Birds." *Oecologia* 109 (1997), 132–141.

H. Craig. "Isotopic Variation in Meteoric Waters." *Science* 133 (1961), 1702–1703.

T. E. Dawson; J. R. Ehleringer. "Streamside Trees That Do Not Use Stream Water." *Nature* 350 (1991), 335–337.

M. J. DeNiro; S. Epstein. "Influence of Diet on the Distribution of Carbon Isotopes in Animals." *Geochimica et Cosmochimica Acta* 42 (1978), 485–506.

———. "Mechanism of Carbon Isotope Fractionation Associated with Lipid Synthesis." *Science* 197 (1977), 261–263.

S. N. Dudd; R. P. Evershed. "Direct Demonstration of Milk as an Element of Archaeological Economies." *Science* 282 (1998), 1478–1481.

H. E. Gove. *From Hiroshima to the Iceman: The Development and Applications of Accelerator Mass Spectrometry.* Bristol, England: Institute of Physics Publishing, 1999.

K. A. Hobson; L. I. Wassenaar. "Stable Isotope Ecology: An Introduction." *Oecologia* 120 (1999), 312–313.

E. G. Johnson; A. O. Nier. "Angular Aberrations in Sector Shaped Electromagnetic Lenses for Focusing Beams of Charged Particles." *Physical Review* 91 (1953), 10–17.

W. Kutschera. "Accelerator Mass Spectrometry: From Nuclear Physics to Dating." *Radiocarbon* 25 (1983), 677–691.

W. F. Libby. "Atmospheric Helium Three and Radium from Cosmic Radiation." *Physical Review* 69 (1946), 671–672.

J. Mattauch; R. Herzog. "Mass Spectrograph." *Zeitschrift für Physik* 89 (1934), 786–795.

D. E. Matthews; J. M. Hayes. "Isotope-Ratio-Monitoring Gas Chromatography–Mass Spectrometry." *Analytical Chemistry* 50 (1978), 1465–1473.

C. R. McKinney et al. "Improvements in Mass Spectrometers for the Measurement of Small Differences in Isotope Abundance Ratios." *Review of Scientific Instruments* 21 (1950), 724–730.

R. A. Muller. "Radioisotope Dating with a Cyclotron." *Science* 196 (1977), 489–494.

B. F. Murphey. "The Temperature Variation of the Thermal Diffusion Factors for Binary Mixtures of Hydrogen, Deuterium, and Helium." *Physical Review* 72 (1947), 834–837.

A. O. Nier. "The Development of a High Resolution Mass Spectrometer: A Reminiscence." *Journal of the American Society for Mass Spectrometry* 2 (1991), 447–452.

———. "A Mass Spectrometer for Isotope and Gas Analysis." *Review of Scientific Instruments* 18 (1947), 398–411.

———. "The Practicality of the Impractical." *Journal of the Minnesota Academy of Sciences* 32 (1964), 12–16.

A. O. Nier; E. A. Gulbransen. "Variations in the Relative Abundance of the Carbon Isotopes." *Journal of the American Chemical Society* 61 (1939), 697–698.

R. Park; S. Epstein. "Carbon Isotope Fractionation during Photosynthesis." *Geochimica et Cosmochimica Acta* 21 (1960), 110–126.

MASS SPECTROMETRY

HISTORY

1920

KDKA, the first radio station in the United States, broadcasts the results of the 1920 presidential election.

Between 150,000 and 200,000 lives are lost in earthquakes in Gansu, China.

Mustafa Kemal Atatürk establishes the modern Turkish state.

————. "Metabolic Fractionation of C13 and C12 in Plants." *Plant Physiology* 36 (1961), 133–138.

J. R. Petit et al. "Climate and Atmospheric History of the Past 420,000 Years from the Vostok Ice Core, Antarctica." *Nature* 399 (1999), 429–436.

T. Preston; N. J. P. Owens. "Interfacing and Automatic Elemental Analyser with an Isotope Ratio Mass Spectrometer: The Potential for Fully Automated Total Nitrogen and Nitrogen-15 Analysis." *Analyst* 108 (1983), 971–977.

J. H. Reynolds. "Alfred Otto Carl Nier." In *Biographical Memoirs of the National Academy of Sciences, 1877–*, Vol. 74, 245–265. Washington, D.C.: National Academy Press, 1998.

S. M. Savin. "The History of the Earth's Surface Temperature during the Past 100 Million Years." *Annual Review of Earth and Planetary Science* 5 (1977), 319–355.

L. I. Wassenaar; K. A. Hobson. "Natal Origins of Migratory Monarch Butterflies at Wintering Colonies in Mexico: New Isotopic Evidence." *Proceedings of the National Academy of Sciences* 95 (1998), 15436–15439.

G. J. Wasserburg. "Isotopic Abundances: Inferences on Solar System and Planetary Evolution." *Earth and Planetary Science Letters* 86 (1987), 129–173.

1921

Frederick Soddy is awarded the Nobel Prize in chemistry for his discovery of isotopes of radioactive elements.

1921

The Mexican civil war ends with the killing of President Venustiano Carranza by supporters of General Alvaro Obregón.

The word *robot* is coined by Czech playwright Karel Capek in his science fiction play *Rossum's Universal Robots (RUR)*.

Sergei Prokofiev performs the premiere of his own *Piano Concerto No. 3* in Chicago.

Chapter **3**

*Where the Rubber
Meets the Road*

THE "RUBBER" THAT ARMIES ROLL ON.
FAULTY INSTRUMENTS MEAN BREAK-DOWNS AT THE FRONT.
ACCURATE INSTRUMENTS KEEP 'EM ROLLING. BRISTOL PRECISION

Left and right: World War II posters emphasized the importance of synthetic rubber and high-octane aviation fuel for America's armed forces. Mass spectrometry played an important role in the development of both materials.

On the eve of America's entry into World War II mass spectrometers were still used primarily by academic physicists to study atomic and molecular structure. These physicists composed the only group of experts with the specialized knowledge and experience needed to construct and operate what was a skill-intensive instrumental technique. Early research in mass spectrometry was primarily concerned with compiling data on the accurate masses and isotopic compositions of the elements. By the 1940s this work was largely complete, except for refinement of the measured values.

After the United States entered World War II in 1941, mass spectrometry moved rapidly from the laboratories of academic physicists into those of industrial chemists. Researchers at American petroleum companies were quick to recognize the unique analytical capabilities of mass spectrometry. They applied this instrumental technique to critical wartime needs, primarily the development of synthetic rubber and high-octane aviation fuels. This chapter examines the origins of industrial mass spectrometry during the last years of the Great Depression and through the tumult of World War II. It also explores crucial innovations in instrument design and the impact of mass spectrometry on organic chemistry during the postwar period.

Mass Spectrometry and the Petroleum Industry

The pioneers of American mass spectrometry—Arthur Dempster, Walker Bleakney, Kenneth Bainbridge, Alfred Nier, and John Tate—worked in academic physics laboratories,

MASS SPECTROMETRY

1922

Francis Aston is awarded the Nobel Prize in chemistry for his discovery of isotopes of "inactive elements."

HISTORY

1922

T. S. Eliot publishes his poem *The Waste Land*.

Frederick Banting (left) and Charles Best of Canada develop insulin injection treatment for diabetes.

Benito Mussolini becomes dictator of Italy.

and their research focused on the analysis of the elements and pure compounds. This early work generated an enormous amount of new knowledge about stable isotopes. In the petroleum industry, however, researchers rarely worked with pure compounds. Crude oil, for example, contained a multitude of hydrocarbons that had to be refined into usable products, such as gasoline, diesel fuel, lubricating oils, and asphalt.

Before cracking techniques were introduced in the 1920s, crude oil was separated into its constituent parts through distillation or fractionation. Conventional distillation analysis was limited by the fact that many different hydrocarbons boil within a very narrow temperature range. Further analysis of the hydrocarbon content of these fractions, such as gasoline, was often carried out using infrared and ultraviolet spectroscopy. Since the absorption spectrum of any molecule is unique, the concentration of a specific substance in a mixture could be identified spectroscopically, provided that it had at least one absorption band at a wavelength at which the other components were transparent. If the chemical complexity of the mixture was increased, however, these techniques provided less useful results. One industry analyst

Electron Ionization

One of the oldest and most useful ways to make sample molecules into ions is to "hit" them with electrons. The earliest electron ionization (EI) source was designed by Arthur Dempster at the University of Chicago. All modern versions of the EI source in use today are based on physicist Walker Bleakney's first design, introduced in 1929. A diagram of a typical EI source is shown below.

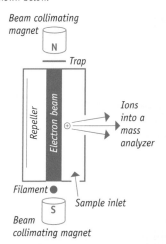

Beam collimating magnet

N

Trap

Repeller

Electron beam

⊕

Ions into a mass analyzer

Filament ●

Sample inlet

S

Beam collimating magnet

The EI source is essentially a box with some holes and a slit in it. The box and all of its parts are held at a fixed voltage. A beam of electrons is injected into the source through holes at opposite sides of the source at right angles to the x-axis of the mass spectrometer. The electron beam comes from an electrically heated filament.

Sample molecules, in the form of a vapor, enter the ion source from the sample inlet system through a separate hole in the source. The interaction of the beam of electrons with sample molecules results in the deposition of energy into the sample molecule. A very small percentage of sample molecules receive so much energy that they end up spitting out an electron, thus forming an ion. Most of these energetic molecules then fragment into two smaller pieces, one of which is charged. Whatever the process, ions are formed from the sample molecules.

Once the ion is formed, a small voltage on the repeller plate gently pushes it toward the exit slit. As soon as the ions reach the exit slit, they are accelerated into the mass analyzer where they are separated.

MASS SPECTROMETRY		1923
		George de Hevesy recognizes the value of isotopes as biological tracers.

HISTORY	1922		1923
	James Joyce publishes *Ulysses*.	The Republic of Ireland gains independence from Britain.	In Germany the Beer Hall Putsch fails, and Adolf Hitler is sentenced to nine months in prison.

noted in 1944 that "chemical and precise fractional distillation analyses are inadequate to provide the high degree of process control required."

In light of these analytical difficulties the mass spectrometer, equipped with an electron ionization source, was especially appealing to petroleum companies for several reasons. First, a mass spectrometer could analyze as many as twenty different components in a gaseous mixture of hydrocarbons. Second, it could do so in only a fraction of the time typically required of existing chemical methods. Finally, and perhaps most important, mass spectrometry was able to distinguish between isomers—two or more molecular species with identical chemical formulas but different molecular structures. Consider, for example, two seven-carbon paraffin isomers, normal heptane and 2,2-dimethylpentane. The mass spectra of the pure

The Mass Spectrum

A mass spectrum can be displayed in many forms. The simplest and most common form is the "stick" plot, such as the one for benzene, shown below. The horizontal axis denotes mass, and the vertical axis denotes relative abundance. As a matter of practice, all

Mass spectrum of benzene
M.W. = 78

mass spectra are shown normalized to the most intense peak, also known as the base peak. Thus all of the smaller peaks in the spectrum are shown relative to the base peak.

The mass spectrum has two regions: the parent, or molecular, ion region and the fragment ion region. Normally, the largest peak in the cluster at the highest mass in the spectrum is considered to be the parent, or molecular, ion. In electron ionization spectra the mass of this ion is equivalent to the mass of the sample molecule with only one electron removed. The smaller peaks just above the molecular ion are referred to as the isotopic cluster. They result from the heavier isotopes in the compound. For example, 1.1 percent of all carbon is ^{13}C. Likewise, about one third of one percent of all nitrogen is ^{15}N. Thus the peaks at one and two mass units higher than the parent ion all have the same elemental composition as the parent ion, but they have heavier isotopes of the constituent elements. Mass spectrometrists use this information to approximate the number of carbon atoms contained in the sample molecule.

There are many other ions whose masses are less than the mass of the molecular ion. These lighter ions are fragments whose masses provide important information about the structure of the molecule. The presence of certain fragments can also signal information about the identity of the sample. For example, spectra showing prominent peaks at masses 31, 45, and 59 daltons (atomic mass units) suggest that the sample compound is probably an alcohol containing at least four carbon atoms.

Wilhelm Wien reviews the current status of research on canal rays in *Zeitschrift für Physik*.

Edwin Hubble confirms the existence of galaxies outside of the Milky Way.

Tokyo and Yokohama are devastated by earthquake and fire, leaving 140,000 dead.

INTERPOL is first organized.

compounds are shown in Figures 1 and 2. Both molecules have the same molecular formula—C_7H_{16}. Since these molecules are structurally different, they will produce different spectra under electron ionization. It was this ability to detect isomers that petroleum chemists found so useful.

Figure 1. *Mass spectrum of normal heptane. Reproduced from Mass Spectral Data, American Petroleum Institute Research Project 44, Serial No. 14 (Washington, D.C.: U.S. Government Printing Office, ca. 1948). Recorded with a CEC Model 21-102 mass spectrometer at the National Bureau of Standards, Mass Spectrometry Laboratory, Washington, D.C., 19 August 1947. Structure of n-heptane and simplified fragmentation pathway.*

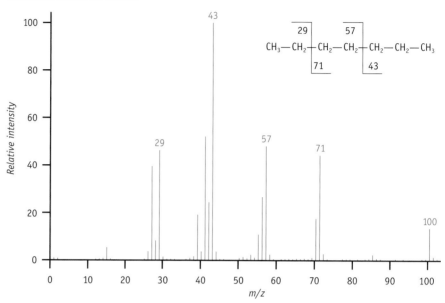

Figure 2. *Mass spectrum of 2,2-dimethylpentane. Reproduced from Mass Spectral Data, American Petroleum Institute Research Project 44, Serial No. 18 (Washington, D.C.: U.S. Government Printing Office, ca. 1948). Recorded with a CEC Model 21-102 mass spectrometer at the National Bureau of Standards, Mass Spectrometry Laboratory, Washington, D.C., 16 July 1947. Structure of 2,2-dimethylpentane and simplified fragmentation pathway.*

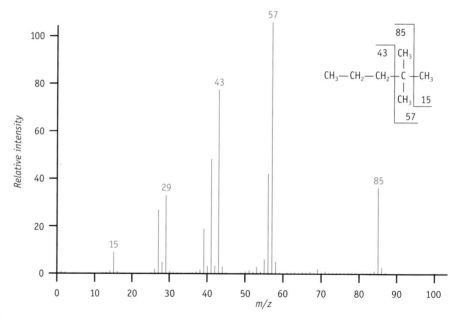

MASS SPECTROMETRY

HISTORY

1924

Westinghouse physicist Vladimir Zworykin develops an early television system.

The first *Australopithecus* fossils are discovered in South Africa by Raymond Dart.

Facial tissue is introduced.

By the end of World War II several industrial laboratories using mass spectrometers had successfully analyzed hydrocarbons containing up to six carbon atoms. In 1948 researchers at the National Bureau of Standards in Washington, D.C., raised the analytical limit even further, to hydrocarbons containing eight carbon atoms. Although mass spectrometers proved extremely valuable for the analysis of such low-molecular-weight hydrocarbon mixtures, the technique was less useful for analyzing mixtures derived from higher-boiling distillates. Two petroleum researchers recalled years later that the problem "was due to the extreme complexity of the mass spectra of the heavier materials, and to the high probability of mutual interference between various isomers as molecular weight increases." After the heptanes (seven carbon atoms), the number of different isomers rises sharply. There are, for example, hundreds of thousands of isomers for the C_{20} family of paraffins.

Rather than try to identify each component in diesel fuels, lubrication oils, and other complex petroleum products, researchers limited their analyses to the total amount of each type of hydrocarbon present in the mixture using matrix algebra techniques. For example, certain types of hydrocarbons share characteristic mass peaks: cycloalkanes at 55, 69, and 83; aromatics, such as the alkylbenzenes, at 91, 105, 119; and paraffins at 43, 57, and 71. Research at several U.S. oil companies facilitated the development of these mass spectral methods for the quantitative analysis of petroleum distillates that provided the analysis in less than one hour. These methods were standardized for general use by the American Society for Testing and Materials (ASTM), and they have provided some of the most useful analytical methods available for high-molecular-weight hydrocarbon analysis. For example, the mass spectrum of a petroleum hydrocarbon mixture is shown in Figure 3. The complexity

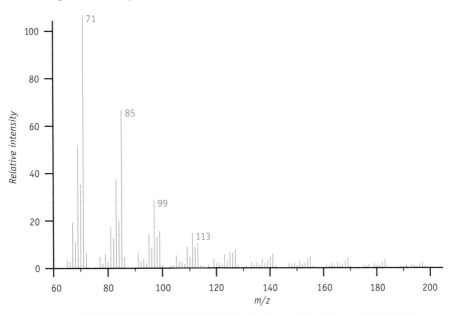

Figure 3. *Mass spectrum of a saturates fraction from a middle distillate recorded with a CEC Model 21-103 mass spectrometer at the National Institute for Petroleum and Energy Research, Bartlesville, Oklahoma, 15 January 1993.*

1925

A German group extends studies done at the University of Danzig on the formation of ozone by cathode rays passing through oxygen.

German physicist E. Rupp produces alkali canal rays by bombarding alkali salts with electrons.

1925

George Gershwin's *Rhapsody in Blue* premiers in New York City.

John T. Scopes is put on trial for teaching the theory of evolution in a Tennessee public school.

Ibn Saud consolidates his control of the country that would later become known as Saudi Arabia.

of the mixture is clearly evident when its mass spectrum is compared with the mass spectra for normal heptane and 2,2-dimethylheptane shown in Figures 1 and 2. Results from the analysis of the mixture using one of the ASTM standard methods are shown in the following table.

Table. Hydrocarbon types analysis of a saturates fraction from a middle distillate by ASTM Method D 2786 (mass spectrum is shown in Figure 3)

Compound type	Volume percentage
Paraffins	61.46
Cycloparaffins	18.00
Bicycloparaffins	10.78
Tricycloparaffins	6.13
Tetracycloparaffins	0.90
Monoaromatics	2.73
	100.00

Commercialization of Mass Spectrometry

The rapid growth of industrial research after World War I created a favorable climate in which commercial mass spectrometry could flourish. Between 1919 and 1936 nearly twelve hundred industrial research laboratories were established by American manufacturing companies. During roughly the same period the number of scientists and engineers working in industry had increased by a factor of ten. Toward the end of this period of unprecedented growth in corporate research and development, the first industrial mass spectrometers were built and sold.

On the eve of America's entry into World War II most practitioners in mass spectrometry used magnetic deflection instruments of the type pioneered by Arthur Dempster at the University of Chicago. These devices, which used massive electromagnets, were bulky and expensive. In 1940 physicist Alfred Nier introduced his new sector mass spectrometer. This instrument, as we saw in chapter 1, used smaller magnets, based on 60- or 90-degree sectors. Both designs were exploited in the first commercial mass spectrometers, one developed by a giant electrical equipment manufacturer in Pittsburgh—Westinghouse Electric and Manufacturing Company—and the other by a little-known commercial instrument company on the West Coast—Consolidated Engineering Corporation (CEC).

Westinghouse Electric entered the field of industrial mass spectrometry in 1941, when physicist John Hipple, one of Walker Bleakney's former graduate students at Princeton University, developed a portable mass spectrometer that used Nier's compact magnetic sector design. Although its initial reception among chemical and petroleum companies was favorable, this portable instrument never sold well. Only about a dozen of Hipple's mass spectrometers were delivered to academic and industrial research laboratories.

MASS SPECTROMETRY

HISTORY	1925	1926	
	The Mid-Atlantic Ridge is discovered, lending support to Wegener's ideas about continental drift.	The first liquid-fueled rocket is successfully launched by Robert Goddard.	Erwin Schrödinger introduces his famous wave equation to describe atomic behavior.

Westinghouse Electric's portable mass spectrometer shown here at the University of Texas in Austin, ca. 1948. This instrument was originally purchased by the Humble Oil and Refining Company for hydrocarbon analysis of petroleum products.

Right: Herbert Hoover, Jr., founder of the United Geophysical Company and its subsidiary, Consolidated Engineering Corporation.

Westinghouse's only major commercial rival in the fledgling mass spectrometer market in 1941 was the Consolidated Engineering Corporation. Founded in 1937 as a wholly owned subsidiary of the United Geophysical Company, CEC manufactured precision instruments for the oil prospecting business. In 1938 Harold Washburn, an electrical engineer recruited from Ernest Lawrence's cyclotron laboratory at Berkeley, began developing a mass spectrometer to identify hydrocarbons in soil. Washburn recognized that measurement of these organic compounds might be useful to geologists looking for petroleum deposits. Unfortunately, Washburn's instrument, which was based on a design developed earlier at the California Institute of Technology, proved unsuitable for petroleum exploration.

In a related move Washburn turned his attention to the development of an instrument that would help solve the analytical problems associated with the production of high-octane aviation fuel for the war effort. By 1940 Washburn and his research group had built a prototype of what was to become CEC's first commercial mass spectrometer—the Model 21-101. Early in 1943, CEC installed the first 21-101 mass spectrometer in the research department of the Atlantic Refining Company near Philadelphia. Unlike the portable magnetic sector device introduced by Westinghouse, the CEC model used Dempster-type geometry, the heart of which was a single-focusing 180-degree mass analyzer. The 21-101 was much larger than its East Coast rival, weighing in at two tons and filling enough space to occupy, as one petroleum industry analyst observed in 1943, "an ordinary-sized kitchen."

Despite its size the CEC 21-101 outperformed older methods of chemical analysis used by researchers in the petroleum industry. In particular, it excelled in the analysis of light hydrocarbon mixtures. The 21-101, for example, could analyze samples eight times faster than

1927	
G. P. Harnwell reports "collisions of the second kind" with the noble gases.	Leonard Loeb clarifies the mechanism by which positive ions can ionize atoms.

1927	
Hirohito becomes Japanese emperor.	Alexander Friedmann and Abbé Georges Lemaître propose the big bang theory.	Civil war breaks out in China between Nationalist and Communist forces.

Researcher with an Atlas-Werke mass spectrometer, shown at the Defence Standards Laboratories in Australia in the early 1960s. This instrument could be used for qualitative organic analysis, quantitative petroleum analysis, and isotope ratio measurements.

Right: Workers assembling General Electric mass spectrometers, 1949.

conventional distillation methods. Whereas distillation often required several liters of a sample for analysis, the mass spectrometer could run a test with only millionths of a liter. The use of an oscillographic recorder provided a dynamic range equal to that of the mass spectrometer, that is, around one part in ten thousand. Moreover, CEC engineers had identified a rubber-based stopcock grease that could be used as a sealant without absorbing the light hydrocarbon gases injected into the sample inlet system.

Armed with the 21-101 and having a firm foothold in the petroleum industry, Consolidated Engineering moved aggressively to exploit the nascent commercial market for industrial mass spectrometers after World War II. Some European manufacturers also made an early entrance into the field, among them Atlas-Werke in Germany and the British firm Metropolitan-Vickers. As Westinghouse withdrew from the instrument business, General Electric (GE) took a bold step into the field. In 1947, GE scientists and engineers built a mass spectrometer based on a modified version of the gas leak detector originally designed by Alfred Nier for the Manhattan Project. Incorporating 60-degree magnetic sector geometry, the GE mass spectrometer matched the CEC 21-101 in terms of sensitivity, resolution, and even cost. The absence of an oscillograph recorder, however, compromised the scan speed and dynamic range of the instrument. GE sold approximately two dozen instruments before exiting the business in 1954.

All of the mass spectrometers manufactured by Consolidated Engineering Corporation used magnetic deflection geometry. Sixteen 21-101 units were sold by 1944 to petroleum

MASS SPECTROMETRY

1928

An early study of the mass spectrum and ionization of water is performed.

HISTORY

1927

Aviator Charles Lindbergh flies solo across the Atlantic Ocean.

Nicola Sacco and Bartolomeo Vanzetti are executed on murder charges of which they are later found to be innocent.

1928

The cartoon short *Steamboat Willie* introduces Mickey Mouse and is the first animated talkie.

companies to monitor refinery streams. During the next twenty years roughly two hundred mass spectrometers from the 21-100 series were sold, ranging in price from $12,000 for the original 21-101 to $40,000 for the more advanced 21-103 models. In addition to a tenfold increase in mass resolving power, these units found widespread application in the petroleum and chemical industries. In 1947, the same year GE introduced its first commercial mass spectrometer, CEC moved into the geochemical and biochemical sciences with the introduction of the 21-201, a Nier-type 60-degree magnetic sector device designed specifically for precision isotope ratio measurements. More than fifty of these "Consolidated-Nier" mass spectrometers, as they were often called, were sold between 1948 and 1954.

The large magnetic deflection mass spectrometers were gradually joined by smaller instruments that incorporated other mass separation technologies. In the late 1950s the Cincinnati Division of the Bendix Aviation Corporation introduced the first commercial time-of-flight (TOF) mass spectrometer. The main advantage of this instrument lay in its ability to make very high-speed—on the order of milliseconds—qualitative identifications of chemical species. During the 1960s the quadrupole mass filter emerged as a viable analytical mass spectrometer suitable for measurements of organic compounds. Applications of TOF and quadrupole mass spectrometry will be discussed in later chapters.

Left: Brochures for Bendix and CEC commercial mass spectrometers.

Right: Harold Washburn (left) and Alfred Nier examine a Consolidated-Nier isotope ratio mass spectrometer.

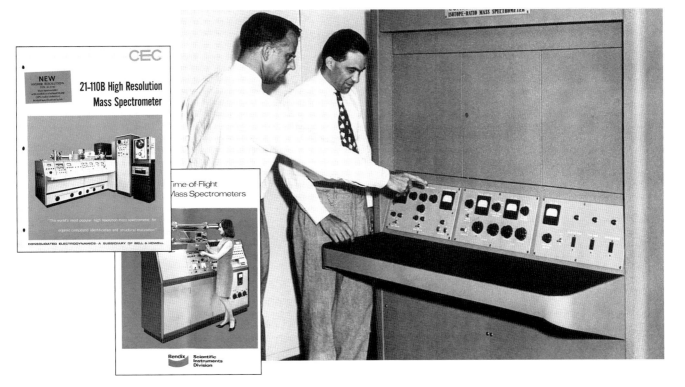

Albert Szent-Györgyi and Charles King independently discover vitamin C.

Ras Tafari becomes the official ruler of Ethiopia, taking the name Haile Selassie I.

Alexander Fleming accidentally discovers penicillin in a mold culture.

Quadrupole Field Mass Filters

The concept of quadrupole mass spectrometry, which uses electrical fields to separate ions based on their charge-to-mass ratio, was introduced in the 1950s by physicist Wolfgang Paul, who shared the 1989 Nobel Prize for physics with Hans Dehmelt for his work.

Paul showed that ions in quadrupolar fields were either stable or unstable depending on their charge-to-mass ratio and the geometrical structure and electrical parameters of the fields. In a linear quadrupole, the fields could be adjusted to allow transmission of an ion with a specific charge-to-mass ratio from one end of the linear quadrupole to the other. Alternatively, a cubic quadrupolar field could be arranged to trap either a single ion or a range of ions, and later the field could be modified to eject the ions in sequence to a detector, as in a quadrupole ion trap. Since in both mass analyzers ions are separated only on the basis of their charge-to-mass ratio, instruments that use this concept act as true mass filters. Time-of-flight and magnetic-sector mass spectrometers separate ions based on other properties of mass, such as energy or momentum.

Mass filters were readily accepted for combination with chromatographs since they had unit mass resolution over a range from 10 to 1,000 daltons and fast scanning capability. These operational parameters are ideal for following the rapidly changing effluents from gas and liquid chromatographs. Modern quadrupole mass filters and ion traps have resolving powers as high as 10,000 and upper mass range limits of 3,000 to 4,000 daltons, but this mass analyzer design is still the basis for low-cost mass spectrometry and tandem mass spectrometry.

Organic Mass Spectrometry

In a spinoff from the development of mass spectrometric methods for monitoring refinery process streams, petroleum researchers and other scientists found new ways to explore the fundamentals of ion formation. This research was motivated in part by the need to understand the factors affecting the fragmentation patterns of molecules under electron ionization. In particular, petroleum researchers were concerned with the effect of ion source temperature and sample pressure on the spectrum. In addition, they wanted to guarantee the reproducibility of fragmentation patterns from instrument to instrument. Another area of interest was investigated by physicist John Hipple and his associates at Westinghouse. Hipple noticed that some mass spectra contained broad or diffuse peaks, usually at nonintegral m/z values. Further research yielded an explanation in 1945. Hipple was able to show that the diffuse peak patterns resulted from the decomposition of ions after they had left the ion source but before they entered the magnetic field of the mass analyzer. On the basis of this observation he concluded that the diffuse peaks corresponded to unstable ions. These "metastable" peaks, as they were now called, provided valuable information about the pathway of a particular molecular decomposition and the structure of the intact molecule itself.

By the 1960s chemists were well versed in the mechanisms of ion decomposition, which helped them establish the mass spectrometer as one of the tools of choice for elucidating the structure of organic compounds. Aside from simple carbon-carbon chain cleavage mechanisms, which give rise directly to fragment ions, rearrangement mechanisms were discovered

MASS SPECTROMETRY	1929

Werner Jacobi investigates multiple charge formation in mercury canal rays.

HISTORY	1929

Al Capone's South Side Chicago crime syndicate carries out the execution-style murders of six members of the North Side Bugs Moran gang in the St. Valentine's Day Massacre.

William Faulkner publishes *The Sound and the Fury*.

Vatican City is granted independence from Italy.

to explain the presence of anomalous fragment ions in the spectrum. A particularly well-known study, conducted by Standard Oil of Indiana chemist Seymour Meyerson in the 1950s, elucidated the mechanism for the formation of a seven-membered carbon ring (the tropylium ion) during the fragmentation of alkylbenzenes. Another example is the rearrangement of a hydrogen atom in a six-membered transition state formed during the fragmentation of ions having a carbonyl group or other group containing a localized charge. Deuterium labeling experiments have shown unequivocally that the hydrogen atom that rearranges comes from the carbon atom third removed from the group with the charge. This mechanism is known as the McLafferty rearrangement.

Studies in organic chemistry diverged into two separate areas: one involved in understanding in greater detail the fundamentals of ion formations and fragmentations, the other in applying the technique to an ever-broadening range of organic compounds. Fundamentals will be discussed in more detail in the next chapter. The applications ranged from the use of high-resolution mass spectrometry to obtain accurate mass measurements of the molecular ions of newly synthesized compounds for elemental composition determinations to the use of gas chromatograph–mass spectrometry techniques to identify natural products, flavors and essences, and environmental pollutants—essentially, any area in which an organic compound was volatile at temperatures used in mass spectrometry.

Time-of-Flight Mass Analyzers

The operating principle behind time-of-flight mass spectrometry (TOF-MS) is really very simple. Ions are formed in the source by a pulse process, either by a switched electron beam or a laser pulse. The ion bunches thus created are accelerated into the analyzer, which is nothing more than a field-free tube. Since all the ions are accelerated by the same voltage and thus have the same energy, their velocity will be a function of their mass. The lighter ions will traverse the tube, from source to detector, more rapidly than the heavier ions. A high-speed multi-channel plate detects the arrival of the ions of different mass at different times, thus creating a spectrum.

The ionization and acceleration process occurs many thousands of times a second, and high-speed computers add together a number of individual spectra to obtain a single spectrum. Note that contrary to most mass analyzers the formation of ions and mass analysis is a pulsed process. And, in principle, the mass range of the TOF-MS is unlimited.

The earliest commercial instrument based on the TOF-MS mass analyzer was marketed by the Bendix Aviation Corporation in the late 1950s. This instrument had limited resolving power and was eventually displaced by the introduction of the quadrupole mass filter. More recently, improvements in electronics, design, and computational resources resulted in the return of the time-of-flight mass analyzer, and it is one of the most popular instruments available today.

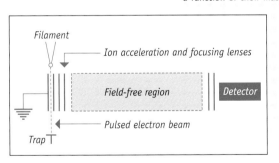

Filament

Ion acceleration and focusing lenses

Field-free region

Detector

Pulsed electron beam

Trap

1930

German physicist E. Ruchardt observes ^{18}O isotope with a parabola mass spectrograph.

1930

The stock market crash on Wall Street heralds the beginning of the Great Depression.

Clarence Birdseye commercializes frozen foods.

Clyde William Tombaugh discovers the planet Pluto.

Suggested Reading

"Analysis of Butadiene and Styrene by Mass Spectrometer." *Journal of the Franklin Institute* 239 (1945), 52.

Annual Book of ASTM Standards. Vol. 05.02. Philadelphia: American Society for Testing and Materials, 1990.

"Better Plane Fuel." *Business Week*, 25 April 1942, 62–64.

J. L. Brady. "The Spectrometric Analysis of Gases." *Oil and Gas Journal* 43 (12 Aug. 1944), 87–90.

A. K. Brewer; V. H. Dibeler. "Mass Spectrometric Analyses of Hydrocarbon and Gas Mixtures." *Journal of Research of the National Bureau of Standards* 35 (1945), 125–139.

R. A. Brown et al. "Mass Spectrometer Analysis of Some Liquid Hydrocarbon Mixtures." *Analytical Chemistry* 20 (1948), 5–9.

R. A. Brown; F. W. Melpolder; W. S. Young. "Analysis from Gases to Waxes by Mass Spectrometry." *Petroleum Processing* 7 (Feb. 1952), 204–211.

J. Delfosse; W. Bleakney. "A Mass Spectrum Analysis of the Products of Ionization by Electron Impact in Nitrogen, Acetylene, Nitric Oxide, Cyanogen and Carbon Monoxide." *Physical Review* 56 (1939), 256–260.

F. H. Field; J. L. Franklin. *Electron Impact Phenomena and the Properties of Gaseous Ions.* New York: Academic Press, 1957.

J. L. Franklin; F. H. Field. "The Energies of Strained Carbonium Ions." *Journal of Chemical Physics* 21 (1953), 550–551.

J. L. Franklin; H. E. Lumpkin. "Lack of Resonance Energy in Gaseous Carbonium Ions." *Journal of Chemical Physics* 19 (1951), 1073–1074.

R. S. Gohlke. "Time-of-Flight Mass Spectrometry and Gas-Liquid Partition Chromatography." *Analytical Chemistry* 31 (1959), 535–541.

J. A. Hipple; E. U. Condon. "Detection of Metastable Ions with the Mass Spectrometer." *Physical Review* 68 (1945), 54–55.

H. C. Hoover, Jr.; H. W. Washburn. "A Preliminary Report on the Application of Mass Spectrometry to Problems in the Petroleum Industry." *Petroleum Technology* 3 (May 1940), 1–7.

———. "Analysis of Hydrocarbon Gas Mixture by Mass Spectrometry." *California Oil World* 34 (1941), 21–22.

D. A. Hounshell. "The Evolution of Industrial Research in the United States." In *Engines of Innovation: U.S. Industrial Research at the End of an Era*, edited by R. S. Rosenbloom and W. J. Spencer, 13–85. Boston: Harvard Business School Press, 1996.

C. M. Judson. "The First Commercial Mass Spectrometer." Unpublished manuscript. Copy preserved at the Chemical Heritage Foundation, Philadelphia.

———. "The GE Mass Spectrometer." Unpublished manuscript. Copy preserved at the Chemical Heritage Foundation, Philadelphia.

A. Langer. "Rearrangement of Peaks Observed in Some Mass Spectra." *Journal of Physical Chemistry* 54 (1950), 618–629.

"Mass Spectrometry: New Quantitative Method for Determining Light Hydrocarbons in Process Control." *National Petroleum News* 35 (2 June 1943), R258–262.

C. A. McDowell, ed. *Mass Spectrometry.* New York: McGraw-Hill, 1963.

F. W. McLafferty. "Mass Spectrometric Analysis: Broad Applicability to Chemical Research." *Analytical Chemistry* 28 (1956), 306–316.

MASS SPECTROMETRY	1931
	E. O. Lawrence invents the cyclotron.

HISTORY	1930		1931
	Mohandas Gandhi leads a nonviolent civil disobedience campaign against the British salt tax in India.	Grant Wood's *American Gothic* is first exhibited at the Art Institute of Chicago.	Wiley Post and Harold Gatty become the first to circumnavigate the earth by airplane.

F. W. McLafferty; M. C. Hamming. "Mechanism of Rearrangements in Mass Spectra." *Chemistry and Industry* 42 (1958), 1366–1367.

S. Meyerson. "Correlations of Alkylbenzene Structures with Mass Spectra." *Applied Spectroscopy* 9 (1955), 120–130.

————. "Reminiscences of the Early Days of Mass Spectrometry in the Petroleum Industry." *Organic Mass Spectrometry* 21 (1986), 197–208.

S. Meyerson; P. N. Rylander. "Organic Ions in the Gas Phase. III. $C_6H_5^+$ Ions from the Benzene Derivatives by Electron Impact." *Journal of the American Chemical Society* 79 (1957), 1058–1061.

J. A. Miller. *Workshop of Engineers: The Story of the General Engineering Laboratory of the General Electric Company, 1895–1952.* Schenectady, N.Y.: General Electric Company, 1953.

D. Milsom. "The Mass Spectrometer and the Control Laboratory." *Petroleum Refiner* 26 (Oct. 1947), 83–89 [719–725].

"New Analytical Techniques Keep Pace with Processes." *National Petroleum News* 36 (2 Feb. 1944), R118.

T. A. Norris; V. P. Nero. "Mass Spectrometry." *Lubrication* 68 (1982), 25–40.

"Recording Mass Spectrometer." *Review of Scientific Instruments* 19 (1948), 283–284.

O. L. Roberts. "Quantitative Analysis by Mass Spectrometry." *Petroleum Engineer* 14 (May 1943), 109–111, 114, 116.

P. N. Rylander; S. Meyerson. "Organic Ions in the Gas Phase. I. The Cationated Cyclopropane Rings." *Journal of the American Chemical Society* 78 (1956), 5799–5802.

P. N. Rylander; S. Meyerson; H. M. Grubb. "Organic Ions in the Gas Phase. II. The Tropylium Ion." *Journal of the American Chemical Society* 79 (1957), 842–846.

P. H. Spitz. *Petrochemicals: The Rise of an Industry.* New York: John Wiley & Sons, 1988.

W. E. Stephens. "A Pulsed Mass Spectrometer with Time Dispersion." *Physical Review* 69 (1946), 691.

J. T. Tate; P. T. Smith; A. L. Vaughan. "Dissociation of Propane, Propylene and Allene by Electron Impact." *Physical Review* 48 (1935), 525–531.

D. D. Taylor. "A Modified Aston Type Mass Spectrometer and Some Preliminary Results." *Physical Review* 47 (1935), 666–667.

H. W. Washburn et al. "Mass Spectrometry." *Industrial and Engineering Chemistry, Analytical Edition* 17 (1945), 74–81.

H. W. Washburn; H. F. Wiley; S. M. Rock. "The Mass Spectrometer as an Analytical Tool." *Industrial and Engineering Chemistry, Analytical Edition* 15 (1943), 541–547.

W. C. Wiley. "Bendix Time-of-Flight Mass Spectrometer." *Science* 124 (1956), 817–820.

————. "Time-of-Flight Mass Spectrometer with Improved Resolution." *Review of Scientific Instruments* 26 (1955), 1150–1157.

W. T. Ziegenhain. "Spectrograph to Analyze Oil Will Feature Laboratory." *Oil and Gas Journal* 40 (14 Aug. 1941), 64.

Direct Quotations

Page 35, "New Analytical Techniques," 1944, p. R118.

Page 37, Norris and Nero, 1982, p. 26.

Page 39, "Mass Spectrometer Analysis of Gasoline," 1943, p. 44.

German researcher R. Dopel observes a charge exchange effect when hydrogen canal rays pass through helium gas.

The Empire State Building is completed in New York City.

Japanese forces invade northern China, conquering the region of Manchuria.

Chapter **4**

Fundamentals

The formation of ions is a critical part of every application of mass spectrometry. Mass analysis, regardless of the kind of analyzer or source involved, requires ions. The sample itself or the ions generated from it must first be vaporized or otherwise placed into the gas phase to be analyzed. Once formed, ions are subject to change and modification before leaving the ionization source. They may break apart spontaneously or in delayed response to an excess of internal energy during the ionization process. They may also fragment as a result of collisions with neutral molecules and undergo further chemical changes. Moreover, such reactions can take place in the mass analyzer, after ions have left the ion source.

The applications of mass spectrometry discussed throughout this volume may appear to take these fundamental processes of ion formation and modification for granted. Nevertheless, it is an inescapable fact that all applications were, and are, dependent on an understanding of the fundamental physical and chemical processes involved in ion formation and fragmentation. The creation over many years of this understanding by researchers in the mass spectrometry community was truly an enabling accomplishment. Without their work, the broad range of mass spectrometric applications would not have been possible, and our knowledge of gas-phase ion chemistry would be nonexistent. The history of inquiry into these processes fundamental to gas-phase ion chemistry—and thus to mass spectrometry as a whole—is the subject of this chapter.

Chemist Henry Eyring, whose contributions to the study of chemical kinetics helped mass spectrometrists understand the nature of ion formation, interaction, and dissociation.

MASS SPECTROMETRY	1932

Harold Urey discovers deuterium.

HISTORY	1932

Carl David Anderson makes the first observation of antimatter.

James Chadwick discovers the neutron.

A group of about 5,000 World War I veterans is attacked by federal troops on the Anacostia Flats in Washington, D.C. The veterans had been demonstrating for early payment of their war bonuses.

The history of research into fundamental gas-phase ion chemistry consists of several long and interconnected threads, starting at the very beginnings of mass spectrometry. The interconnection and length of this history place a comprehensive, chronological treatment beyond the limitations of this volume. This chapter focuses on those aspects of fundamental gas-phase ion chemistry most salient to mass spectrometry, taking a thematic look at the early studies of ion formation, fragmentation, and interactions.

Overview

These three types of studies can be best understood by viewing them all through the common lens of "energetics." Ion formation has to do with the energy required to produce an ion from a neutral molecule. Ion fragmentation is concerned with the energy required to break particular chemical bonds that hold the ion together. Ion interactions involve the energies associated with encounters between ions and neutral molecules, other ions, or surfaces.

The nature and power of this energetics view are perhaps best illustrated by the following example. Investigations of ion-molecule reactions enabled researchers to determine a quantity—"proton affinity." In turn, the proton affinity is a critical factor in the development of recent ionization techniques, such as chemical, electrospray, and matrix-assisted laser desorption ionization. The energetics revealed in studies of the fundamentals of ion interactions gave knowledge of a quantity that, in turn, proved critical in the development of new techniques for ion formation.

Experimental Approaches

Two experimental approaches have been used in virtually all mass spectrometric studies of fundamental ion chemistry. Sometimes they have been employed alone and sometimes in combination. The simpler of the two is the study of threshold, or onset, processes. In a threshold process the minimum energy required to form an ion from a molecule is measured. While threshold processes are quite simple conceptually, achieving an accurate measurement of these energies is experimentally challenging. The technique of examining threshold energies, and energies for the onset of particular phenomena, has been used to study a wide array of problems in fundamental ion chemistry: gaseous ion energetics, energy transfer during the formation of ionized and neutral molecules, and bond dissociation energies of diatomic and polyatomic molecules. In the classical version of this approach researchers studied ions that were formed from intact molecules in the gas phase that had been directly introduced into the ion source of the mass spectrometer. More recently the threshold approach has been used to study the energetics of collision-induced fragmentations. As with so many other aspects of mass spectrometry, classical approaches are continually revisited and modified to yield new, significant results.

The second experimental approach used for studies of fundamental ion chemistry focuses on the kinetics of ion reactions and interactions. Up to the 1950s the emphasis of

MASS SPECTROMETRY	1933	
	The theory of angular focusing for sector magnets is published by English physicist N. F. Barber.	

HISTORY	1932	1933	
	Fascist dictator Antonio de Oliviera Salazar comes to power in Portugal.	The Tennessee Valley Authority begins supplying electricity to many parts of the southeastern United States.	Prohibition is repealed with the ratification of the 21st Amendment to the U.S. Constitution.

Fundamentals

these kinetic studies was on simple reactions between atomic ions and small neutral molecules. Throughout the 1950s and 1960s the emphasis switched to the study of the interactions of more complex ions and molecules, particularly hydrocarbons. Until the 1960s all such studies were limited to molecules and ions whose thermal energies were established by the temperature of the ion source. Thus, one could control the energies over a very limited range by adjusting the ion source temperature. Once tandem mass spectrometry was developed in the 1960s, experiments involving a much wider range of controllable ion energies—from a few electron volts to kiloelectron volts—could be conducted in collision cells within a mass analyzer.

Indeed, the effects of these experimental approaches to fundamental ion chemistry are manifested through the development of new ionization techniques. Chemical ionization—in which molecules of a reagent gas react with an analyte, ionizing it for mass analysis—originated in fundamental kinetic studies of ion-molecule reactions; so too did the technology of atmospheric pressure ionization. The use of proton-transfer reactions to form specific ions has also been a direct outgrowth of kinetic studies.

Chemical Ionization

The chemical ionization (CI) source was one of the first departures from electron ionization (EI). In the diagram below you can see that the CI source is not very different from the EI source discussed earlier. The main differences are that the holes in the source are smaller, and there is an extra hole for the "reagent gas" to enter.

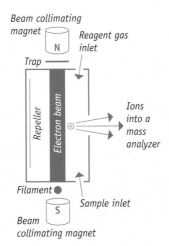

Beam collimating magnet

N

Reagent gas inlet

Trap

Repeller

Electron beam

Ions into a mass analyzer

Filament ●

Sample inlet

S

Beam collimating magnet

The reagent gas is usually a small molecule, such as methane or water. The pressure of the reagent gas is about a thousand times greater than the pressure in the ion source during EI operation. In addition, the electron current flowing through the CI source is also greater. This increase creates gaseous ions from the reagent gas. When the sample enters the CI ion source, it reacts with the reagent gas ions to form new ions. Just as in the electron ionization source, once the new ion is formed, the repeller gently pushes the ion toward the exit slit so that it can be accelerated into the mass analyzer.

Why is this ionization technique used, and why is it so important? CI is a gentler ionization technique than electron ionization. Since much less energy is imparted to the sample molecule in converting it to an ion, less ion fragmentation occurs. This means that for some compound classes, such as large biological molecules, an adduct parent ion is observed where no parent ion would be observed under the more destructive conditions of electron ionization. It also forms the basis for a whole field of research known today as gas-phase ion chemistry.

President Franklin Roosevelt's New Deal programs are instituted to relieve the effects of the Great Depression.

Polyethylene is invented by Reginald Gibson and E. W. Fawcett.

Adolf Hitler comes to power in Germany.

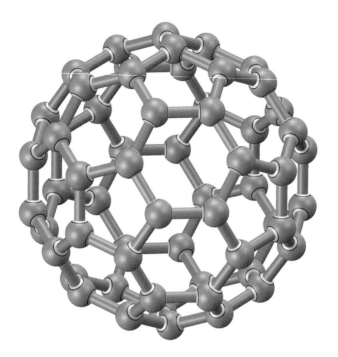

Fullerenes are extremely durable carbon-based molecules.

Furthermore, these fundamental investigations have provided insight into ion interactions in the condensed, liquid, or solid phase—reactions that are not open to investigation by other analytical techniques. For example, the equilibrium ion-molecule studies pioneered by chemists Frank Field and Burnaby Munson and their collaborators, as well as by chemist Paul Kebarle and his co-workers, have led to a better understanding of the energy involved when an ion goes into solution, such as an ion dissolving in water. In addition, the discovery of fullerenes and other carbon-caged structures is an outgrowth of fundamental kinetic studies of cluster formation.

Like the analytical applications of mass spectrometry, the investigation of fundamental ion chemistry has depended on the development of new technologies and new instrumentation. Before the development of adequate vacuum systems, for example, the study of ion-molecule reactions was extremely difficult because source and analyzer pressures could not be controlled independently. Today, the availability of high-performance differentially pumped vacuum systems has made such studies routine in laboratories throughout the world.

Ion Formation

Mass spectrometers of the 1920s and 1930s were designed and built by physicists for studies of the physical nature of small molecules. Experimental studies focused on the testing of theories, especially those connected to the then-new quantum mechanics. Many experiments were aimed at extending observations made previously in optical spectroscopy into mass spectrometry. In these optical spectroscopy experiments, emission-line appearances, intensities, and structure were studied as functions of energy and pressure. Mass spectrometrists carried this practice from optical spectroscopy into their experimental tests of theory. They developed an understanding of mass spectral lines and their appearances, intensities, and structures by varying ionization energies and gas pressures. Furthermore, these researchers gained knowledge about the ion chemistry resulting from these changes. Experiments of this type resulted in a more detailed knowledge of the chemistry of the reactions under investigation and thus were an improvement over optical methods. These early mass spectrometric experiments are exemplified by a group of seminal papers published in the *Physical Review*

MASS SPECTROMETRY

1934

The double-focusing mass spectrograph is developed by Josef Mattauch and Richard Herzog.

HISTORY

1934

A general labor strike grips San Francisco.

1934–35

Chinese Communist forces embark on the "Long March," fleeing to remote regions of northwestern China to regroup in the face of pressure by Nationalist forces under Chiang Kai-shek. During this period Mao Zedong emerges as the leader of the Chinese Communists.

in the mid-1920s. These papers demonstrated that a mass spectrometer could be used to great effect in the direct measurement of the appearance potentials of specific molecules.

Mass spectrometry research in the field of ion-formation energetics remained at a rather low but constant level until World War II. During the next twenty years the field grew rapidly, especially as the commercial market for mass spectrometers began to expand. Moreover, the industrial applications of mass spectrometry had also provided an important impetus to the field's rapid postwar growth. As we saw in chapter 3, the petroleum industry's use of mass spectrometers required close attention to the ionization processes taking place in the ion sources of its instruments. Ionization potentials, appearance potentials of fragment ions, and bond dissociation energies of virtually all the light hydrocarbons—C_1 to C_6—were determined via mass spectrometry by researchers in the petroleum industry. A classic monograph summarizing this fundamental ion chemistry research was written jointly by two former presidents of the American Society for Mass Spectrometry. The analytical methods presented in this treatise were subsequently applied to molecules other than hydrocarbons. Molecules from a large number of organic, inorganic, and organometallic compounds were studied in academic, industrial, and government laboratories. Today, applications of the threshold approach are found in studies of ion-molecule reactions. Collision-induced dissociations, carried out under conditions in which the energy of the reactants is carefully controlled, have been used to measure bond energetics of both covalently and noncovalently bound complex ions.

Ion Fragmentation

The dissociation of ions in the course of energetic collisions was evident in the early work of J. J. Thomson and Francis Aston. The gases studied in their instruments were ionized by electrical discharges, the precursor to modern electron ionization methods. Although revealed in their experiments, collision-induced dissociation was not studied in its own right until much later. Studies of appearance potentials and the dissociation of ions at higher energies became more sophisticated in the 1920s and 1930s, when researchers at Princeton University and the University of Vienna used more advanced mass spectrometers to investigate ion behavior.

Mass spectral studies of low-energy electron-induced fragmentations of simple polyatomic molecules were conducted during the 1930s and early 1940s by a number of groups. For example, the fragmentation of acetylene to H^+, C^+, CH^+, C_2^+, and C_2H^+ was studied by physicist John Tate and his coworkers at the University of Minnesota in 1935. Starting in the 1940s, researchers in petroleum laboratories developed a more extensive understanding of more complex molecules. These low-energy fragmentation studies led to the development of rules for the interpretation of the mass spectra of organic compounds. As we saw in chapter 3, two noteworthy achievements of this era were associated with ion rearrangement reactions: the formation of the tropylium ion in alkylbenzenes and the McLafferty rearrangement.

The spontaneous dissociation of previously activated ions was first studied in the latter part of the 1940s. The kinetic energy released in these processes was recognized as the source of the width of metastable peaks observed in sector mass spectrometers after the ions had left the ion source. Measurements of metastable peaks were used to characterize dissociation pathways of energetic ions. In addition, accurate determination of the kinetic energy release was used to correct threshold thermochemical data.

By the 1950s ion fragmentation studies were highly advanced. Canadian researchers, for example, were among the first to recognize that high-energy collisions involve two steps: a collisional activation step, in which energy is distributed among the various internal vibrational and rotational modes of the ion, and a delayed dissociation step in which some fraction of the internal energy is released in the translational modes. Meanwhile, researchers in the United States performed elegant work with multi-sector instruments on high energy ion-target gas and ion-foil collisions. The cross-sections of the energy transfer processes as a function of collision energy were a special focus of European researchers.

Ion Interactions

Interactions between ions and neutral molecules were first observed by J. J. Thomson, when he identified H_3^+ in his parabola mass spectrograph. This pioneering work was later discussed by Francis Aston in his landmark treatise, *Isotopes,* published in 1924. Further study and an explicit description of ion-neutral molecule collision processes were made in the 1920s by Princeton physicist Henry Smyth.

During this same period collision-induced reactions of the "second kind" were also reported. These reactions of the second kind are known today as second-order reactions, meaning that they require two components present in approximately equal concentrations. An example of such a reaction is shown here, in which charge transfer between an argon ion and a neutral oxygen molecule occurs:

$$Ar^+ + O_2 \rightarrow O_2^+ + Ar$$

Unlike research on ion formation and fragmentation, experimental studies of ion-molecule interactions declined during the 1930s and 1940s. Instead, researchers focused their efforts on theoretical aspects of these processes. In the mid-1930s, for example, researchers, including such distinguished scientists as Henry Eyring and Michael Polanyi, carried out important theoretical investigations on this topic.

In the 1950s experimental studies of ion interactions experienced a resurgence. The field expanded rapidly into studies of chemical reactions in flames, radiochemistry, atmospheric chemistry, and flowing afterglows. For mass spectrometry itself, this work had its most profound impact on the development of new ionization techniques. Chemical ionization—that is, the reaction of reagent ions with the molecules of a sample—is grounded in a necessary understanding of ion-molecule interactions. Equally significant is the recognition that chemical ionization produced even-electron ions, requiring less energy than odd-electron

MASS SPECTROMETRY	1936	
	Secondary ion mass spectrometry is introduced.	A. J. Dempster describes spark source ionization.

HISTORY	1936	
	In Spain, civil war erupts that will last for three years and bring dictator Francisco Franco to power.	Daniel Bovet of Switzerland discovers the antibacterial properties of sulfanilamide, the first sulfa drug.

53

Fundamentals

Physicist Henry Smyth helped transform Princeton University into a major center for mass spectrometry research during the 1930s.

ions produced by electron ionization. Thus, there is also less fragmentation from the use of chemical ionization and other proton-transfer ionization processes—such as electrospray and matrix-assisted laser desorption.

In the 1960s the dependence of ion-molecule interactions on energy was increasingly explored. The development of tandem-sector mass spectrometers—that is, two or more mass spectrometers coupled together—led to the elucidation of the mechanisms of collisionally induced dissociations. High-energy studies were conducted with sector instruments. The introduction of triple-quadrupole mass spectrometers made it possible to extend this type of research to low-energy collisions. The introduction of the ion-trap mass analyzer, either ion cyclotron resonance (ICR) or quadrupole, led to mass spectrometric experiments in which one could follow the sequential decomposition of an ion through multiple stages. The thermochemistry of these processes was also examined, resulting in an important rule of thumb. Exothermic reactions—reactions that give off heat—proceed at a rate similar to that of collisionally induced dissociation, while endothermic reactions—those that absorb heat—proceed significantly more slowly. This work laid the foundation for subsequent studies of gas-phase acidity and basicity.

Clearly, studies of the fundamentals of ion formation and fragmentation have a rich history that has paid dividends, not only in mass spectrometry but also in other areas of chemistry. While fundamental studies of ions may be the oldest field of study in mass spectrometry, it remains alive and well, with prospects for continued discoveries into the future.

Suggested Reading

L. R. Anders et al. "Ion-Cyclotron Double Resonance." *Journal of Chemical Physics* 45 (1966), 1062–1063.

N. Aristov; P. Armentrout. "Reaction Mechanisms and Thermochemistry of $V^+ + C_2H_2p$ (p = 1-3)." *Journal of the American Chemical Society* 108 (1986), 1806–1819.

F. W. Aston. *Isotopes.* London: Edward Arnold, 1924.

P. Ausloos; S. Lias. "Gas-Phase Radiolysis of Propane." *Journal of Chemical Physics* 36 (1962), 3163–3170.

H. Barton. "The Ionization of HCl by Electron Impacts." *Physical Review* 30 (1927), 614–633.

———. "Single and Double Ionization of Argon by Electron Impacts." *Physical Review* 25 (1925), 469–483.

H. Barton; J. Bartlett. "The Positive Ray Analysis of Water Vapor Ionized by Impact of Slow Electrons." *Physical Review* 31 (1928), 822–826.

1937

Researchers at Columbia University use mass spectrometry to investigate isotopes of hydrogen in organic compounds.

1937

The British Broadcasting Company airs its first television broadcasts.

Pure DNA is first isolated by Andrei Nikolaevitch Belozersky.

The airship *Hindenburg* burns in Lakehurst, New Jersey, ending the age of lighter-than-air travel.

Amelia Earhart disappears over the Pacific Ocean during an attempted round-the-world flight.

Fundamentals

J. S. Brodbelt-Lustig; R. G. Cooks. "Dissociation of Protonated 2,2,6-Trimethylcyclohexanone and Related Compounds upon High and Low Energy Collisional Activation." *International Journal of Mass Spectrometry and Ion Processes* 86 (1988), 253–272.

R. B. Cody et al. "Consecutive Collision-Induced Dissociations in Fourier Transform Mass Spectrometry." *Analytical Chemistry* 54 (1982), 2225–2228.

P. Dawson. "The Collision Induced Dissociation of Protonated Water Clusters Studied Using a Triple Quadrupole." *International Journal of Mass Spectrometry and Ion Processes* 43 (1982), 195–209.

H. Eyring; J. Hirschfelder; H. Taylor. "The Theoretical Treatment of Chemical Reactions Produced by Ionization Processes, Part I: The Ortho-Para Hydrogen Conversion by Alpha Particles." *Journal of Chemical Physics* 4 (1936), 479–491.

E. Ferguson et al. "Laboratory Studies of Helium Loss Processes of Interest in the Ionosphere." *Planetary and Space Science* 12 (1964), 1169–1171.

F. Field; J. Franklin. *Electron Impact Phenomena and the Properties of Gaseous Ions.* New York: Academic Press, 1957.

———. "Electron Impact Studies of Some Aromatic Hydrocarbons: Implications Regarding Certain Aromatic Reactions." *Journal of Chemical Physics* 22 (1954), 1895–1904.

F. Field; M. Munson. "Reactions of Gaseous Ions. XIV. Mass Spectrometric Studies of Methane at Pressures of 2 Torr." *Journal of the American Chemical Society* 87 (1965), 3289–3294.

W. Fite et al. "Ion-Neutral Collisions in Afterglows." *Discussion of the Faraday Society,* no. 33, 264–272.

J. Franklin; F. Field. "The Resonance Energies of Certain Organic Free Radicals and Ions." *Journal of the American Chemical Society* 75 (1953), 2819–2821.

G. L. Glish; S. A. McLuckey; K. G. Asano. "Determination of Daughter Ion Formulas by Multiple Stages of Mass Spectrometry." *Journal of the American Society for Mass Spectrometry* 1 (1990), 166–173.

W. B. Hanson. "Upper-Atmosphere Helium Ions." *Journal of Geophysical Research* 67 (1962), 183–188.

G. Harnwell. "Ionization by Collisions of the Second Kind in the Rare Gases." *Physical Review* 29 (1927), 683–692.

J. A. Hipple; E. U. Condon. "Detection of Metastable Ions with the Mass Spectrometer." *Physical Review* 68 (1945), 54–55.

T. Hogness; R. Harkness. "The Ionization Process of Iodine Interpreted by the Mass Spectrograph." *Physical Review* 32 (1928), 784–790.

R. E. Kaiser, Jr., et al. "Collisionally Activated Dissociation of Peptides Using a Quadrupole Ion-Trap Mass Spectrometer." *Rapid Communications in Mass Spectrometry* 4 (1990), 30–33.

P. Kebarle; E. Godbole. "Mass-Spectrometric Study of Ions from the Alpha Particle Irradiation of Gases of Near Atmospheric Pressures." *Journal of Chemical Physics* 39 (1963), 1131–1132.

P. Kebarle et al. "The Solvation of the Hydrogen Ion by Water Molecules in the Gas Phase: Heats and Entropies of Solvation of Individual Reactions, $H + (H_2O)n-1 + H_2O \rightarrow H + (H_2O)n$." *Journal of the American Chemical Society* 89 (1967), 6393–6399.

P. Knewstubb; T. Sugden. "Spectrometric Studies of Ionization in Flames. I. The Spectrometer and Its Applications to Ionization in Hydrogen." *Proceedings of the Royal Society of London, Part A* A255 (1960), 520–537.

MASS SPECTROMETRY

HISTORY

1937

A steelworkers' strike in Chicago is quelled when police fire on the crowd of strikers, leaving ten dead in the infamous Memorial Day Massacre.

Pablo Picasso paints *Guernica* in remembrance of the brutal attack on Basque civilians in northern Spain by the German Luftwaffe in support of Franco's fascists.

P. Knewstubb; A. Tickner. "Mass Spectrometry of Ions in Glow Discharges. III. Nitrogen and Its Mixtures with Hydrogen and Oxygen." *Journal of Chemical Physics* 37 (1962), 2941–2999.

F. Lampe. "The Direct Radiolysis and the Radiation-Sensitized Hydrogenation of Ethylene." *Radiation Research* 10 (1959), 691–701.

E. Lindholm. "Ionization and Fragmentation of Nitrogen by Bombardment with Atomic Ions. Dissociation Energy of Nitrogen." *Arkhiv Fysik* 8 (1954), 257–264.

T. Magnera et al. "Production of Hydrated Metal Ions by FIB or FAB Sputtering: Collision-Induced Dissociation and Successive Hydration Energies of Gaseous Cu with 1-4 Water Molecules." *Journal of the American Chemical Society* 111 (1989), 5036–5043.

C. E. Melton; P. S. Rudolph. "Mass Spectrum of Acetylene Produced by 5.1 MeV Alpha Particles." *Journal of Chemical Physics* 30 (1959), 847–848.

B. D. Nourse; R. G. Cooks. "Aspects of Recent Developments in Ion-Trap Mass Spectrometry." *Analytica Chimica Acta* 228 (1990), 1–21.

R. Ogg; M. Polanyi. "Mechanism of Ionic Reactions." *Transactions of the Faraday Society* 31 (1935), 604–620.

H. Rosenstock; C. Melton. "Metastable Transitions and Collision-Induced Dissociations in Mass Spectra." *Journal of Chemical Physics* 26 (1957), 314–322.

H. Rosenstock et al. "Energetics of Gaseous Ions." *Journal of Physical and Chemical Reference Data* 66 (1977), Supp. 1.

H. Smyth. "Primary and Secondary Products of Ionization in Hydrogen." *Physical Review* 25 (1925), 452–468.

H. Smyth; E. Stueckelberg. "Ionization by Collisions of the Second Kind in Mixtures of Oxygen with the Rare Gases." *Physical Review* 32 (1928), 779–783.

V. L. Talrose; A. K. Lyubimova. "Secondary Processes in the Ion Source of the Mass Spectrograph." *Doklady Akademii Nauk SSSR* 86 (1952), 909–912.

J. Tate; J. Smith; A. Vaughan. "A Mass Spectrum Analysis of the Products of Ionization by Electron Impact in Nitrogen and Acetylene." *Physical Review* 43 (1933), 1054.

J. J. Thomson. *Rays of Positive Electricity and Their Application to Chemical Analysis.* London: Longmans, Green, 1913.

1938

Walker Bleakney and John Hipple describe the cycloidal mass analyzer.

Consolidated Engineering Corporation is formed.

1938

Howard Florey and Ernest Chain first isolate the penicillin that Alexander Fleming had identified twelve years earlier as having antibiotic properties.

Orson Welles's radioplay adaptation of *War of the Worlds* causes mass hysteria on the eastern seaboard of the United States as many mistake it for a news broadcast of a real invasion from Mars.

Thought to have been extinct for 70 million years, a live coelacanth fish is caught in the ocean near South Africa.

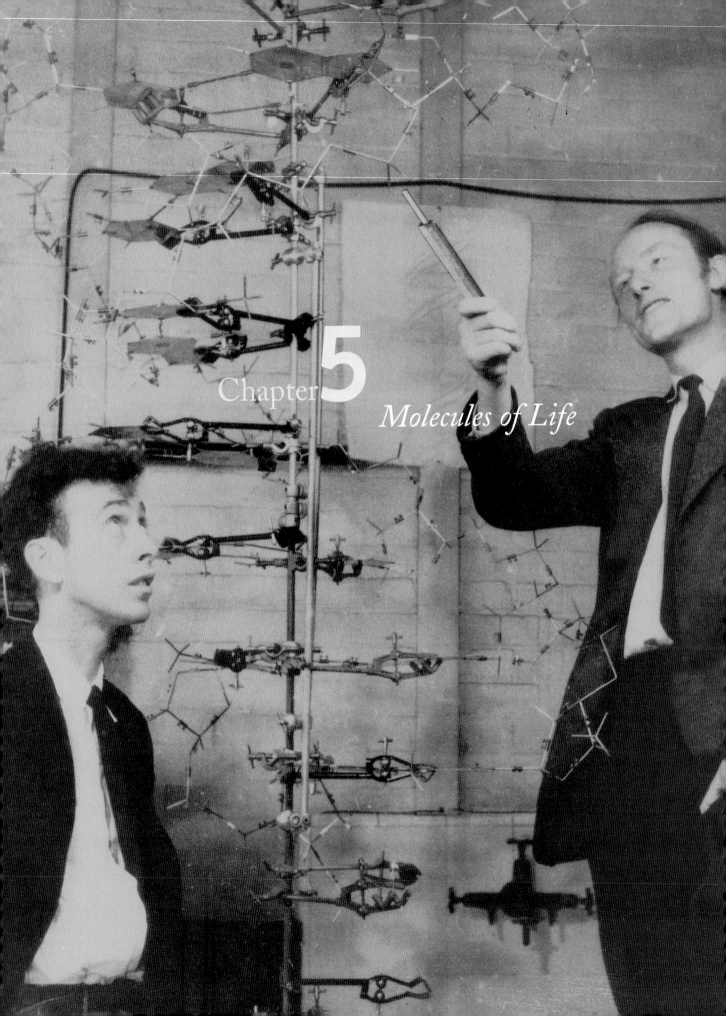

Chapter **5**
Molecules of Life

Right: Model of modified human insulin dimer displaying tertiary structure. Portions with arrows are beta sheets. Portions with coils are alpha helices. Hydrogen-deuterium exchange studies of this molecule by mass spectrometry provide additional insights into its function.

Left: James Watson and Francis Crick examine a model of the DNA double helix.

What are the biochemical processes involved in living cells? How do biological molecules communicate with one another to produce healthy cells? In what fashion are these processes impaired in diseased cells? Scientists have been seeking the answers to these fundamental questions for decades and will continue to do so in the new millennium. Although there is still much more to learn, scientists have made significant advances, ranging from the discovery of the structure of DNA in 1953 to the elucidation of the human genome half a century later. Throughout this period scientists have used new concepts and analytical instruments to unravel the mysteries of life. One of the most important analytical instruments employed in this quest has been, and continues to be, the mass spectrometer.

Throughout the twentieth century, scientists have used mass spectrometry to study increasingly larger molecules, such as those associated with the processes of life. During the 1930s researchers began using the isotopes of carbon, nitrogen, oxygen, and hydrogen as "tracers" to reveal biological processes in living organisms.

Using mass spectrometry over the last two decades, chemists and biologists have greatly expanded biochemistry, analyzing the structure and function of more complex combinations of molecules. In the same way, mass spectrometry has become a critical enabling technology in structural biology—the study of proteins, carbohydrates, cell membranes, and other large biological molecules. Through its profound impact on both biochemistry and structural

MASS SPECTROMETRY	1939	
	Variation of isotope ratios in nature is first demonstrated at Harvard University.	Nitrogen isotopes are first used in protein metabolism studies.

HISTORY	1938	1939	
	Germany erupts in officially approved violence and vandalism against the Jewish community on 9 November in what becomes known as *Kristallnacht*.	Germany invades Poland, thus beginning World War II.	The Spanish Civil War ends with the victory of the fascists under Francisco Franco.

Atomic Mass Unit vs. Dalton

The unified atomic mass unit (u) is defined as the mass of one atom of ^{12}C divided by 12. Its approximate physical value is 1.66×10^{-27} kilograms. This term is the standard measure of mass used in the atomic, molecular, and nuclear sciences. The dalton, named after the famed English chemist John Dalton, is equivalent to (u), although it is most often used by mass spectrometrists working in the biochemical sciences.

biology, mass spectrometry has opened up new vistas for our understanding of life and for our capacity to improve human health.

More recent technological innovations, such as liquid chromatography mass spectrometry (LC-MS) and new "soft ionization" techniques, have overcome these limitations. While some of these instrumental developments will be discussed here, most of the more recent improvements in ionization, such as fast atom bombardment, electrospray ionization, and matrix-assisted laser desorption ionization, will be treated in the discussion of structural biology in the second half of this chapter. It is in the analysis of the largest biomolecules, often weighing more than 100,000 daltons, that structural biology is concerned and that these sophisticated ionization methods have found widespread use. The demand for analytical instrumentation for research of this type has grown dramatically in recent years.

Biochemistry and Mass Spectrometry before World War II

At the University of California during the 1930s physicist Ernest Lawrence was using his cyclotron to produce radioactive tracers for medical research, especially for the therapeutic treatment of deadly diseases. Meanwhile, on the East Coast, chemist Harold Urey and his colleagues at Columbia University used stable isotope tracers to start a different, though broadly related, research tradition, one based on the use of mass spectrometry to explore new biochemical processes and functions.

At a meeting of the American Physical Society in New Orleans in 1931, Urey announced his discovery of deuterium (^2H), one of the heavy isotopes of hydrogen. Stable isotopes like deuterium were very useful to researchers investigating the dynamic processes of metabolic pathways and other biological phenomena. A particular advantage of these stable isotopes is that they did not expose the organism to radiation, thus minimizing potential health risks to researchers and their subjects. They could be followed through complex biochemical processes with repeated measurements with a mass spectrometer. As never before, biochemists had the power to explore processes in living organisms with the utmost sensitivity and accuracy.

Among Urey's early experiments with deuterium were studies that explored the characteristics of ^2H_2O, also known as "heavy water." When he had enough heavy water on hand, Urey encouraged other researchers at Columbia to investigate the compound and its uses.

MASS SPECTROMETRY	1939		
	E. O. Lawrence receives the Nobel Prize in physics for the invention of the cyclotron.		

HISTORY	1939		
	China's Yangtze River floods, killing between 500,000 and one million people.	Lise Meitner recognizes the results of experiments by Otto Hahn as nuclear fission.	The World's Fair in New York City gives visitors a look at the world of tomorrow.

Molecules of Life

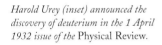

Second Series   *April 1, 1932*   Vol. 40, No. 1

THE

PHYSICAL REVIEW

A HYDROGEN ISOTOPE OF MASS 2 AND ITS CONCENTRATION*

By Harold C. Urey, F. G. Brickwedde, and G. M. Murphy**

Columbia University and the Bureau of Standards

(Received February 16, 1932)

Abstract

In a recent paper Birge and Menzel pointed out that if hydrogen had an isotope with mass number two present to the extent of one part in 4500, it would explain the discrepancy which exists between the atomic weights of hydrogen as determined chemically and with the mass spectrograph, when reduced to the same standard. Systematic arrangements of atomic nuclei require the existence of isotopes of hydrogen H^2 and H^3 and helium He^5 to give them a completed appearance when they are extrapolated to the limit of nuclei with small proton and electron numbers. An isotope of hydrogen with mass number two has been found present to the extent of one part in about 4000 in ordinary hydrogen; no evidence for H^3 was obtained. The vapor pressures of pure crystals containing only a single species of the isotopic molecules H^1H^1, H^1H^2, H^1H^3 were calculated after postulating: (1) that the rotational and vibrational energies of the molecules are the same in the solid and gaseous states; (2) that in the Debye theory of the solid state, the θ's are inversely proportional to the square roots of the molecular masses; (3) that the free energy of the gas is given by the free energy equation of an ideal monatomic gas; and (4) that there is a zero point lattice energy equal to $(9/8)R\theta$ per mole. The calculated vapor pressures of the three isotopic ... es in equilibrium with their solids at the triple point for ordinary hydro- ... in the ratio $p_{11}:p_{12}:p_{13} = 1:0.37:0.29$. The isotope was concentrated in three ... of gas by evaporating large quantities of liquid hydrogen and collecting the ... ch evaporated from the last two or three cc. Sample I was collected from the ... tion of six liters evaporated at atmospheric pressure and samples II and III ... r liters, each, evaporated at a pressure only a few millimeters above the triple

... ese samples and ordinary hydrogen were investigated for the visible, atomic ... series spectra of H^2 and H^3 from a hydrogen discharge tube run in the condi- ... orable for the enhancement of the atomic spectrum and for the repression of ... ecular spectrum, using the second order of a 21 foot grating with a dispersion of ... er mm. When with ordinary hydrogen, the times of exposure required to just re- ... strong H^1 lines were increased 4000 times, very faint lines appeared at the cal- ... positions for the H^2 lines accompanying $H^1\beta$, $H^1\gamma$ and $H^1\delta$ on the short wave- ... ide and separated from them by between 1 and 2A. These lines do not agree in ... ngth with any known molecular lines and they do not appear on the plates ... ith the discharge tube operating under conditions favorable for the production ... ong molecular spectrum and the repression of the atomic spectrum. With or- ... tation Approved by the Director of the Bureau of Standards of the U. S. Depart- ... merce.

..., Urey and G. M. Murphy, Columbia University, F. G. Brickwedde, Bureau of

1

Harold Urey (inset) announced the discovery of deuterium in the 1 April 1932 issue of the Physical Review.

One of Urey's colleagues was chemist Hans Clarke, who used heavy water to synthesize deuterium-labeled organic molecules. Chemist David Rittenberg and physician Rudolf Schoenheimer, also prompted by Urey, joined Clarke's group. Schoenheimer, who received an appointment to Columbia in 1933 after fleeing Nazi Germany, suggested that the group use deuterium labels to trace cholesterol metabolism. This concept served as the centerpiece of a novel approach because stable isotopic tracers had never been used for biochemical studies of metabolism.

In 1935 Schoenheimer and Rittenberg used deuterium-labeled fats to explore energy-transfer processes in animals. Scientists had long recognized the end products of animal metabolism, but little was known about the pathways linking the synthesis and degradation of carbohydrate, fat, and protein molecules into carbon dioxide, water, and urea. Prevailing theories held that fat deposits were simply reserve supplies, only used by the body when diet was inadequate. Schoenheimer and Rittenberg labeled the fatty acids in linseed oil by hydrogenating it with deuterium and fed the labeled fatty acids to mice. Their results showed that nearly half of the deuterium-labeled acids had been transferred to the fat deposits in each mouse. This experiment, based on mass spectrometric analysis of the deuterated tracer, confirmed that fatty acids are routinely stored as fatty deposits and are just as regularly drawn upon. Schoenheimer and Rittenberg also used deuterium tracers and mass spectrometry in their investigations of carbohydrates and amino acids, further developing our view of the body as a site of constant activity, with its many constituents always undergoing dynamic change.

These early studies by Schoenheimer and Rittenberg led to a series of seminal papers that probed a wide range of biochemical pathways. Their work using isotopic nitrogen was among the earliest efforts to apply mass spectrometry to a biochemical problem. "Our aim has been to use [^{15}N] as an indicator for the investigation of protein and amino acid metabolism in healthy animals on ordinary diets in a way similar to that in which deuterium had been used

---

1940

A. O. Nier and John Dunning determine that ^{235}U is the fissionable isotope of uranium.

A. O. Nier develops the single-focusing mass spectrometer using a 60-degree magnetic sector.

Westinghouse Electric begins to develop a portable mass spectrometer for commercial sale.

1940

Germany invades Norway, Denmark, France, and the Netherlands.

Russian revolutionary Leon Trotsky is assassinated in Mexico City.

1941

Germany invades the Soviet Union.

The Grand Coulee Dam opens in Washington State.

Left: Rudolf Schoenheimer.

Right: George de Hevesy.

for the study of fat and sterol metabolism," Schoenheimer and Rittenberg explained in the first of two pioneering papers that they published in 1939. Whereas radioactive isotopes had already been used by future Nobel laureate George de Hevesy in studies of phosphate metabolism, only the stable isotopes of carbon and hydrogen could be used at this time to study biological processes. Since oxygen and nitrogen have no radioactive isotopes, mass spectrometry was used to make tracer studies with these elements possible. Meanwhile, similar biochemical experiments using ^{13}C provided by Alfred Nier were conducted in 1940 by bacteriologist Harland Wood at Iowa State University.

Schoenheimer and Rittenberg exploited the high sensitivity of mass spectrometry to investigate minute changes in the nitrogen isotope ratio as the tracer passed through the organism. Conventional methods of measuring atomic weight were simply too time consuming, required large samples, and were not particularly sensitive to small variations in isotopic composition. "High sensitivity and accuracy of the analytical procedure are required," they insisted. Today, researchers still use stable tracers and isotope ratio mass spectrometry to measure human metabolic processes. Determination of whole-body energy metabolism using doubly labeled water ($^{2}H_2^{18}O$) and studies of fatty acid metabolism using labels of ^{13}C and ^{2}H are just two of the more well-known examples.

While pioneering scientists such as Schoenheimer and Rittenberg at Columbia and Wood at Iowa State were using stable isotopes to explore biochemical processes, mass spec-

MASS SPECTROMETRY

1942

E. O. Lawrence leads development of the "Calutron" preparative scale mass spectrometer for the separation of uranium isotopes.

HISTORY

1941

Orson Welles directs and stars in *Citizen Kane,* considered by many to be the best film of all time.

The United States enters World War II after the Japanese attack on Pearl Harbor.

1942

Enrico Fermi conducts the first controlled nuclear chain reaction at the University of Chicago.

trometers were still complex, specialized instruments rarely found outside of academic research laboratories. No commercial mass spectrometers were yet available. In fact, the instrument that Schoenheimer and Rittenberg used was a single-focusing mass spectrometer whose operation was based on a design previously introduced by Walker Bleakney. Although commercial mass spectrometers had become available in the 1940s, these instruments were used primarily in petroleum laboratories, largely because they were so expensive. For the most part, homemade mass spectrometers were used by biochemists well into the 1950s.

In the 1950s gas chromatography was able to separate a limited class of biomolecules— volatile, thermally stable compounds, such as some fatty acids, steroids, and derivatives of amino acids and carbohydrates. Gas chromatograph mass spectrometers (GC-MSs) were incapable of analyzing molecules weighing more than several hundred daltons. To extend the powers of the GC-MS beyond these limitations, technological innovation was a must. Extension of the mass range of biomolecules that could be analyzed with the GC-MS was significantly advanced by the development of new coupling technologies for marrying GC with MS. The GC-MS innovation favored by biochemists was the jet separator developed

Gas Chromatography Mass Spectrometry

A gas chromatograph separates the components of a complex mixture. The chromatograph consists of a column—a tube usually several millimeters in diameter and several meters long—coated with a liquid phase. The mixture of analytes is injected into the column and swept through it by a carrier gas, usually helium. The constituent molecules pass through the column at different rates. At the column exit the various components of the mixture are separated from one another and emerge as pure compounds. After separation the emerging pure compounds are introduced one after the other into the mass spectrometer. This hybrid instrument, first demonstrated in the late 1950s, has played a decisive role in the analysis of such highly complex mixtures as petroleum-based hydrocarbons and organic and small biological molecules.

The gas chromatograph and the mass spectrometer, however, are not completely compatible. The instruments operate at opposite pressure extremes. The outlet of the gas chromatograph is at atmospheric pressure, while the ion source of the mass spectrometer operates at a pressure of just a few billionths of an atmosphere. Pressure-reducing interfaces, called molecular separators, were originally used to remove the chromatographic carrier gas, while allowing most of the analyte to continue through the interface to the ion source. Gas chromatographic column technology, especially the development of narrow bore capillary columns, has improved to the point where carrier gas-flow rates are ten times lower now than when the first interfaces were introduced. Moreover, today's mass spectrometers are equipped with differentially pumped vacuum systems capable of handling the total mass flow of chromatographic effluent, thus permitting direct coupling of the two instruments.

Mass spectra are continuously recorded throughout the duration of the chromatographic run and stored in computer memory. Mass assignment and intensity measurements are performed on the fly, and spectra can be submitted to a library search in real time while the chromatographic run is in progress. In a typical GC-MS run, lasting anywhere between thirty minutes and one hour, many thousands of mass spectra can be recorded.

1943

Consolidated Engineering Corporation (CEC) installs its first commercial mass spectrometer (Model 21-101) at Atlantic Refining Company in Philadelphia.

1943

120,000 Japanese-Americans are interred in detention camps in the western United States.

The United States defeats a Japanese fleet in the Battle of Midway, turning the tide of the Pacific War.

The Nazis are defeated by the Soviets at the horrific Battle of Stalingrad, turning the tide of war on the Eastern Front.

by Swedish chemist Ragnar Ryhage in the early 1960s. The jet separator was successful in biochemical analyses because of the minimum exposure of intact molecules to hot surfaces, thereby ensuring that biomolecules did not degrade in the interface.

In the 1960s, Klaus Biemann, a chemist at the Massachusetts Institute of Technology, realized that improvements in the technologies of GC-MS had made it a logical, but as yet unexploited, analytical tool for studies of amino acid sequences in proteins. These sequences were key to understanding the structure and function of proteins. Prior to mass spectrometric analysis, protein sequencing required some tedious and time-consuming tasks. Large proteins had to be broken into smaller peptides and separated. After separation, each peptide then had to be derivatized, or chemically modified, to increase volatility. Only then was there even a possibility that each peptide could be analyzed by GC-MS to determine its individual amino acid sequence. Of course, GC-MS did not become a universally accepted analytical tool for peptide sequencing overnight. Biemann's primary contribution was to recognize that the technique would become a superior analytical tool over older wet chemical techniques. Subsequent improvements in GC-MS instrumentation, along with new soft ionization techniques—chemical ionization, thermospray ionization, and field desorption, among them—made this transition possible.

Applications of GC-MS to Other Areas of Biochemistry

Enzymes catalyze many of the biological reactions supporting life. These enzymatic processes, in turn, are subject to carefully balanced regulatory mechanisms within the body's cells. The concentration of metabolites—chemical compounds that are created in the course of such biological processes—in the cells of a normal, healthy organism is fairly constant. In cases where a cell, or class of cells, is defective or diseased, the concentration of metabolites may change markedly. They either decrease or increase, and entirely new abnormal metabolites may appear. By the early 1970s researchers were using GC-MS to explore the link between the presence of disease and changes in the concentration of metabolites in human cells. Researchers working in the Institute of Clinical Biochemistry at the University of Oslo, for example, acknowledged in 1973 that "the number of metabolites occurring in [blood and urine] samples . . . is likely to exceed several thousand, and no analytical system presently exists for detecting them all."

The Oslo group then conducted GC-MS clinical screenings of eight hundred patients. Results of these analyses yielded fresh information on more than forty known disorders and also provided valuable data on the metabolic profiles of cells in other diseased states. This analytical approach to clinical screening was so productive that four new diseases were isolated and identified. GC-MS proved most successful in cases in which the changes in the metabolite concentration were relatively large. Mapping changes caused by diseased states, such as liver and kidney disease and cancer, rather than congenital defects of the metabolism proved far more difficult. In such cases researchers only observed small changes in metabolic

An uprising of Polish Jews confined in the Warsaw ghetto is brutally crushed by German forces.

Rodgers and Hammerstein's *Oklahoma!* opens on Broadway.

Scuba (self-contained underwater breathing apparatus) is invented by Jacques Cousteau and Emil Gagnan.

profiles. The ability to detect small changes, even trace changes, with GC-MS was enhanced by integrating computer data-handling systems with these instruments. In the early 1970s this technology was in its infancy. Another limiting factor was the lack of standard reference spectra for biological compounds and drug metabolites within library databases. Dramatic improvements in data acquisition, handling, and screening of critical biological molecules occurred during the next quarter century owing to the advance of computer technology.

In addition to investigating inborn errors of metabolism and diseases, mass spectrometry has also benefited researchers examining other features of human metabolism. Two fields of research in particular stand out: the analysis of stable isotopes used to trace internal biochemical pathways and the determination of rates of production and transformation for compounds involved with biochemical processes. The mass spectrometer has given biochemists a powerful tool for learning more about the complex processes involved in the maintenance of health and the diagnosis and treatment of disease.

Other applications of mass spectrometry to clinical screening have involved the analysis of small, isotopically labeled molecules. Doubly labeled water ($^2H_2^{18}O$), for example, has been used to determine whole-body energy metabolisms during studies of normal exercise and obesity in humans. Such studies have given clinicians insight into metabolic disorder–induced obesity. Similarly, the analysis of carbon dioxide produced by the breakdown of isotope-labeled precursor compounds has been used to measure rates of metabolism of sugar and fat malabsorption and to diagnose ulcers and liver dysfunction. Metabolic pathways involved in the biosynthesis of endogenous molecules in humans, such as steroids and sugars, have also been discovered by mass spectrometry. The use of stable isotopic labels that could be followed by mass spectrometry was preferable to the use of radioactive tracers. Such tracers were a safety concern, particularly for children and women of child-bearing age.

Mass Spectrometry and Structural Biology

Only in the last twenty years have mass spectrometry–based studies become significant in the field of structural biology. The origins of this transformation can be traced back to major innovations in analytical instrumentation, again particularly to the development of soft ionization techniques. Two major obstacles had prevented the widespread use of mass spectrometry in research on massive biological molecules. First, the analyte, or sample material, had to be volatile. Most biological molecules are fragile, heat sensitive, or polar, making them unsuitable for ionization by older methods. Derivatization methods—chemical modifications of the analyte—were sometimes used to increase volatility and direct the course of fragmentation, but these techniques were often less than satisfactory because the derivatized sample typically increased in molecular weight and chemical complexity. Equally problematic was the limited mass range of the mass spectrometers available before the 1980s. Molecules with masses larger than 1,000 daltons were difficult to analyze in any meaningful way, especially in cases in which an extremely small amount of sample material was available. Although

1944

A CEC 21-101 users group forms in Pasadena, California.

The U.S. National Bureau of Standards makes fifteen hydrocarbons available as calibration standards for mass spectrometers.

Albert Hoffman of Switzerland discovers LSD to be a powerful hallucinogen.

1944

Allied armies invade Normandy on D-Day.

The International Monetary Fund and the World Bank are created.

Where Does the Muscle Go?

A common observation in aging individuals is that they tend to lose muscle protein, which contributes to physical frailty. This phenomenon is observed in other clinical conditions (HIV/AIDS, kidney failure, diabetes, trauma, cancer, sepsis, burns, prolonged exposure to the microgravity of space flight, and prolonged bed rest). High-precision isotope ratio mass spectrometry plays a key role in investigating the underlying mechanisms responsible for muscle protein wasting and in determining appropriate intervention strategies.

Stable isotope tracer methodologies combined with mass spectrometric detection are used to measure the in vivo rate of incorporation of amino acids into skeletal muscle proteins, a measure of muscle protein synthesis rate. Substrates, such as carbohydrates, lipids, and amino acids, labeled with stable nuclides of C, N, O, and H are administered to the participant. The metabolic fate of these substrates is "traced" by measuring the abundance of the stable-labeled compound in blood, urine, and muscle and other tissue specimens obtained during the experiment. The isotope abundance in these specimens can be measured using high-precision isotope ratio mass spectrometry.

From experiments conducted at the Washington University School of Medicine, it was determined that frail elderly and HIV-infected individuals have lower-than-normal rates of muscle protein synthesis. As a leading researcher describes it, "The amount of muscle in the body is like the amount of money we have in a bank account: it is a function of how fast we deposit and withdraw it. When dysregulation occurs, the rates for these processes change and, along with it, the amount of muscle in our 'muscle account.'"

This information helps researchers identify genes and proteins that are under- or over-expressed in wasting muscles. In combination with mass spectrometric measures of muscle protein synthesis rate, researchers are now investigating the role of a novel growth factor, myostatin, that has been shown to regulate muscle growth in cattle.

efforts were under way in individual laboratories to overcome these limitations, most biological macromolecules remained elusive for the mass spectrometrist.

By the 1970s improvements in analytical instrumentation began to overcome these formidable obstacles to analyzing biological macromolecules, like proteins, by mass spectrometry. The introduction of plasma desorption mass spectrometry (PDMS) enabled researchers to ionize polar, nonvolatile molecules from a *solid* matrix without the aid of chemical derivatization. PDMS used the energy of fission particles from californium (^{252}Cf) to desorb and ionize sample molecules bound to a thin foil. When coupled to a time-of-flight mass analyzer, spectra could be obtained from molecules weighing tens of thousands of daltons, even at extremely low sample concentrations. PDMS, like most of the ionization methods introduced later, was most effective in the analysis of peptides and proteins. Although it has been largely replaced by other methods, PDMS showed conclusively that polar macromolecules, such as proteins, could be desorbed intact and survive transit through a mass analyzer.

In 1981 fast atom bombardment (FAB) ionization was introduced, allowing such highly polar molecules as peptides and small proteins to be analyzed with a *liquid* matrix. Like PDMS, FAB has been routinely used to ionize molecules weighing more than 10,000 daltons. FAB is also effective for analysis of a wide variety of biological molecules, including

MASS SPECTROMETRY	1945
	CEC introduces the Model 30-103 analog computer for analysis of mass spectra of hydrocarbon mixtures.

HISTORY	1944		1945
	Oswald Theodore Avery shows that DNA carries genetic information.	Women in newly liberated France gain the right to vote.	Franklin Roosevelt, Winston Churchill, and Joseph Stalin meet at Yalta in the Soviet Union to determine the fate of postwar Europe.

Molecules of Life

Protein Structure

Proteins are among the most intensively studied molecular species in the field of structural biology. Protein structure can be broken down into four categories. Primary structure is concerned with the sequence of amino acids that make up the protein; it controls the formation of secondary structure. Secondary structure deals with the local folding of the protein chain, that is, the formation of subunits of the protein into alpha helices or beta sheets, as well as disulfide links between cysteine residues. Tertiary structure covers global folding of the whole protein and controls binding with proteins and other compounds. Finally, quaternary structure addresses the interactions between folded proteins—how two proteins come together as subunits to form a larger protein complex. Each of these levels of structure controls the next higher level of structure.

Protein structure is important because it controls the biological function of the protein. It is particularly important for drug development. Once the protein sequence and structure is known, drugs can be designed to bind with that protein in a certain way to produce a specific therapeutic effect.

peptides and proteins, small oligonucleotides, oligosaccharides, and lipids. Researchers adopted this technology quickly not only because of these intrinsic capabilities but also because FAB ionization sources could be retrofitted with relative ease onto existing instruments. Indeed, the quest for efficient peptide sequencing and the emergence of mass spectrometry in at least one area of what is now called "proteomics" had its start with FAB analysis. Subsequent years saw the introduction of several types of continuous-flow FAB interfaces, permitting the coupling of flowing liquid streams with mass spectrometry. This allowed liquid chromatography and flow injection to be used as a sample introduction method for biological analysis.

The next major innovative ionization technique developed in the 1980s was electrospray ionization (ESI). It was quickly embraced by the biological mass spectrometry community. ESI is extremely sensitive and has been improved to the point at which it has already been used to produce a mass spectrum from a sample containing fewer than one hundred million molecules. Moreover, molecular weights of relatively large proteins can be measured with an accuracy of fifty parts per million. Like FAB, the electrospray ionization source can be retrofitted to many existing mass spectrometers, and since it is a liquid flow device, it is compatible with LC-MS and other liquid flow inlets. Unlike FAB, however, ESI does not require a separate liquid matrix, which simplifies its interface with the mass spectrometer. The principal difference between ESI and FAB is the ability of ESI to produce multiply charged ions. It is this feature that allows mass spectrometers with limited mass ranges to analyze extremely large biological molecules.

Although electrospray ionization was introduced in 1968, only after development in the 1980s was success gained in the analysis of a wide variety of biological molecules, ranging from low-molecular-weight metabolites, lipids, and xenobiotics to high-molecular-weight species

John Hipple and Edward Condon
determine the source of "metastable"
peaks in a mass spectrum.

World War II ends.

The United Nations is founded.

A patent is issued for the first microwave oven.

Fast Atom Bombardment Ionization

Fast atom bombardment (FAB) was introduced in 1981 by researchers in the United Kingdom. This ionization method marked a radical departure from existing techniques, such as electron and chemical ionization. The diagram below shows the essential components of an FAB ion source.

The sample is dissolved in a liquid matrix, usually a very viscous compound with a low vapor pressure, such as glycerol, or what is commonly known as

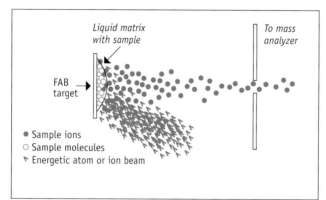

FAB target →

Liquid matrix with sample

To mass analyzer

● Sample ions
○ Sample molecules
▾ Energetic atom or ion beam

glycerine. The mixture is placed on a target at the end of the FAB probe, which is then introduced into the FAB source through a vacuum lock. Once the target is in place in the FAB source, a high-energy beam of neutralized atoms or ions is directed at the target by an FAB gun. Typically, the beam has energies in the range of 6,000 to 10,000 electron volts and is made up of heavy noble gases, such as krypton or xenon. The interaction of the energetic beam with the sample in the liquid matrix results in the formation of sample ions. Frequently, positively charged ions are produced. Since the FAB target is immersed in an electric field, ions leave the surface of the matrix and are immediately accelerated into the mass analyzer.

Fast atom bombardment is ideally suited for ionization of complex organic compounds that are not easily vaporized. Since FAB imparts little energy to the sample molecules, ions can be readily formed from extremely delicate compounds. This advantage is significant for biological applications. Mass spectrometers using FAB technology can analyze peptides, proteins, and other nonvolatile, labile molecules.

like oligonucleotides and oligosaccharides. More recently, low flow-rate variations on this ionization technology, such as microelectrospray and nanospray, have proved capable of excellent performance with remarkably high sensitivities.

Soon after electrospray ionization appeared, matrix-assisted laser desorption ionization (MALDI) was introduced. This technique required no liquid phase and could ionize molecules beyond 300,000 daltons, such as proteins. When coupled to modern time-of-flight mass analyzers, mass measurement accuracies of several parts in ten thousand could be achieved. MALDI was also readily embraced by the biological research community because of its extraordinary ease of use and high-molecular-weight measurement capability. As with other ionization methods, MALDI was most useful for analyzing peptides and proteins, although other biological macromolecules could also be analyzed.

Structure and Function of Biological Molecules

Understanding the structure and function of biological molecules and their integrated roles in health and disease is one of today's foremost research fields. Advances in the technologies

MASS SPECTROMETRY | **1946**

The first description of the time-of-flight mass spectrometer is published in the *Physical Review*.

Metropolitan Vickers produces the first MS-1 mass spectrometer.

HISTORY | **1946**

The Philippines gains independence from the United States.

Nazi leaders are tried in Nuremberg, Germany, for crimes against humanity.

Electrospray Ionization

Electrospray ionization has dramatically extended the ability to apply mass spectrometry to problems of biological significance. It has facilitated the efficient interfacing of liquid chromatography with mass spectrometry, and because it creates a range of multiply charged ions from peptide and protein samples, it greatly extends the mass range of existing instruments.

The sample to be analyzed (the analyte) is mixed with a solvent and sprayed from a narrow tube. Positively charged droplets in the spray move toward the mass spectrometer sampling orifice under the influence of electrostatic forces and pressure differentials. As the droplets move toward the orifice, the solvent evaporates, resulting in an increased surface charge. This increased charge, in turn, causes the droplets to rupture. A succession of these processes yields the analytically useful ions at the sampling orifice.

The principal advantage of electrospray ionization is that large molecules become multiply charged, primarily by the addition of multiple protons. The important aspect of this feature is that a series of peaks appears in the spectrum, representing a distribution of such charges. Each peak represents an independent determination of the molecular weight of the analyte.

This concept is illustrated in the electrospray mass spectrum of Cytochrome C. The peak at *m/z* 825 represents the protein molecule with fifteen added protons, picked up from the spray solution during the electrospray process. The peak at *m/z* 728 is the same protein molecule with seventeen added protons. Using a simple set of linear equations, the mass of the protein can be determined, assuming that each of the peaks in the spectrum differs from the others by a single charge. More elegant mathematical methods can be applied to the deconvolution of the multiple charge states to yield the actual mass of the analyte. Thus, even though the mass spectrometer may have an upper *m/z* limit of 2,000 daltons, it can be used to determine the mass of proteins tens of times larger.

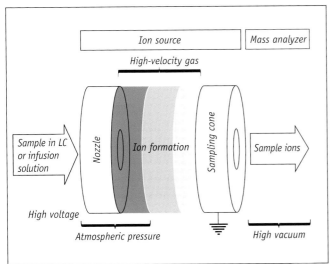

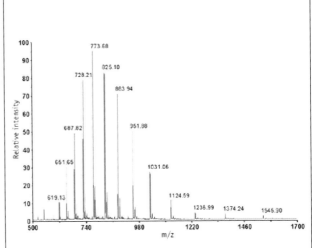

1947
General Electric begins manufacturing commercial mass spectrometers.

1947
Presper Eckert and John Mauchly build ENIAC, the world's first all–electronic computer.

Juan Perón comes to power in Argentina.

Jean-Paul Sartre heralds many existentialist ideas in his brochure *Existentialism and Humanitarianism*.

Matrix-Assisted Laser Desorption Ionization

Matrix-assisted laser desorption ionization (MALDI) was introduced in 1987. This technique has been used primarily to ionize extremely massive biological molecules that often weigh more than 100,000 daltons. Unlike electron-impact ionization methods, which can fragment delicate molecules, MALDI is a nondestructive, or "soft," ionization technique.

MALDI uses laser technology to ionize compounds. Before the development of MALDI, experiments used lasers to vaporize and ionize solid materials and metals. Organic matrices have since been employed to enhance the ionization of biological molecules. The matrix is mixed with the sample and deposited in the form of a small spot on the sample plate. After the solvent evaporates, a crystalline mixture, consisting of the sample and the matrix, is all that remains on the plate. Once the plate is moved into the ion source of the mass spectrometer, a laser beam is focused on the sample. The beam, chosen to have a wavelength that is readily absorbed by the matrix, vaporizes and ionizes both the matrix and the analyte molecules. The resulting ions are immediately accelerated to a constant energy and introduced into the mass analyzer, typically a time-of-flight (TOF) mass spectrometer. Normally, a sample plate can be loaded with as many as a hundred sample spots, making MALDI-TOF a very efficient instrument for analyzing a large number of biological samples.

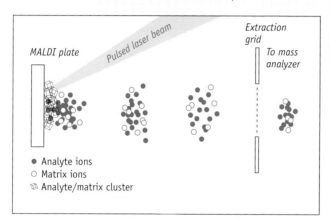

- Analyte ions
- Matrix ions
- Analyte/matrix cluster

of mass spectrometry have had an enormous impact on this field. Innovations in performance—sensitivity, mass range, and speed of analysis—have brought the power of the mass spectrometer to bear on biological molecules. Moreover, innovations in design and data systems have made mass spectrometers much easier to use. Thus, the instruments have moved beyond the hands of traditional mass spectrometrists and into the hands of other scientists also engaged in the pursuit of structural biological knowledge.

Technological advances in mass spectrometry have affected protein and peptide analysis far more than any other area of structural biology. Sequence analysis of peptides up to about 2,000 daltons is routinely accomplished. When combined with a liquid chromatograph, the mass spectrometer became a powerful tool for analysis of protease digests of large peptides and proteins. Mass measurements of peptides and determinations of their sequences allowed researchers to work backward to identify the original protein. This result could be considered one of the most important applications of mass spectrometry and, further, has changed the way biologists approach protein identification.

During the 1990s this area of protein research came to be known as "proteomics." Compilation of large protein databases derived from computer-based translations of DNA

MASS SPECTROMETRY   **1947**

The U.S. National Bureau of Standards and the American Petroleum Institute initiate the distribution of reference mass spectra and the creation of a library of mass spectra.

CEC introduces its "Consolidated-Nier" isotope ratio mass spectrometer.

HISTORY   **1947**

India and Pakistan gain independence from Britain as a result of a long nonviolent campaign led by Mohandas Gandhi.

Chuck Yeager becomes the first person to travel faster than sound in the Bell X-1 experimental aircraft.

Willard Frank Libby develops radiocarbon dating as a method of determining the age of remains of living organisms.

sequences of human genes has enabled researchers to deduce the primary structures of proteins. However, researchers recognized that simple molecular-weight measurements by mass spectrometry could not effectively identify proteins because of modifications that occurred after they were expressed by genes. Nevertheless, mass spectrometrists realized that at least some fragments of proteins would not be modified at all or only to a limited and easily identified extent. The application of mass spectrometry to proteomics was accelerated in the mid-1990s when new computational approaches were developed for rapid protein identification. Combined with large protein databases, accurate mass measurements have become a powerful analytical tool for the proteomic era.

Genomics and proteomics are intimately bound. Genes, coded in DNA, create the proteins that make life possible. Not only has mass spectrometry been devoted to the study of these proteins, but it has also been used to explore DNA. MALDI and ESI have been used successfully to analyze oligonucleotides, the basic structural units of DNA and other nucleic acids. The high sensitivity of ESI allowed researchers to obtain sequence information on oligonucleotides containing more than fifty base pairs with ion-trap, triple-quadrupole, and Fourier transform ion cyclotron resonance (FTICR) instruments. Computer algorithms improved data analysis by providing an efficient means for interpreting the extensive fragmentation generated by collision-induced dissociation of the oligonucleotides. Unlike radiolabeling methods, mass spectrometry provides faster analysis of oligonucleotides with much less sample material. Currently, high-throughput mass spectrometry can produce numerous gene sequences within a single day.

The potential of mass spectrometry for structural biology goes far beyond macromolecular identification: it allows scientists to explore the functional aspects of biomolecules, commonly referred to as their "tertiary structure." Post-translational modifications of proteins —that is, changes after they are created by DNA—are important determinants of the protein's eventual function, and these modifications are especially amenable to characterization by modern mass spectrometry. Sites at which these structural transformations occur can be readily identified by tandem mass spectrometry and then correlated to protein function.

Mass spectrometry has also revealed important information about the role of carbohydrates in the functions of cells and their interactions with viruses. Carbohydrates are important to cellular and virus function through their role in the biochemistry of cell attachment, infectivity, secretion, and solubility and in protection against proteolytic and chemical degradation. Improvements in MALDI and ESI methods have enabled investigators to identify "glycosylation" sites at which carbohydrates are attached. Even the highly complex glycan structures have also been identified.

Mass spectrometry has played an important role in the study of noncovalent interactions, such as intermolecular or intramolecular hydrogen bonding between biomolecules. The potential to study protein-ligand and protein-protein interactions with high molecular specificity has opened new avenues of research for the understanding of biological complexes

---

1948

Researchers at the University of Minnesota design a dual inlet with a changeover valve for rapid sample switching in high-precision isotope ratio mass spectrometry.

Ion cyclotron resonance mass spectrometry (Omegatron) is developed.

1948

The transistor is invented by William Shockley, Walter Brittain, and John Bardeen at Bell Telephone Laboratories.

The Berlin Airlift saves West Berliners from starvation during a Soviet blockade.

In Calcutta, Albanian-born Mother Teresa founds the Order of the Missionaries of Charity to assist the city's poorest residents.

and processes. By using reagents that selectively react with specific amino-acid side chains or nucleotide bases, mass spectrometrists have obtained tertiary and quaternary structural information from the spectra of these complexes. Again, mass spectrometry has given researchers a revealing window into the biomolecular world, allowing them to watch in detail the relationships between structure and function of biomolecules and complexes of interacting biomolecules.

Mass spectrometry has continued to provide knowledge of the structure of proteins, those massive biomolecules folded upon themselves in complex and shifting patterns. Mass spectrometrists first studied these conformational changes by measuring changes in ionic charge states. Proteins exhibit a distinct difference in their electrospray charge-state distribution, dependent on their solution conformers. For example, two charge-state distributions can often be observed for a protein's native (less charge) and denatured (more charge) conformation. This shift in distribution is associated with the additional protonation sites available on the denatured form of the protein. Because native proteins have fewer accessible sites for protonation, the conformation of the protein is reflected in the charge-state distribution. More recently, hydrogen-deuterium (^{1}H-^{2}H) exchange experiments have been employed in this area, using both ESI and MALDI.

Novel areas of recent interest in the application of mass spectrometry to structural biology include immunology, tissue mapping, and huge macromolecular complexes, such as intact viruses. Moreover, the recent completion of the human genome project has pushed the proteome to the forefront of biological research. Many modern mass spectrometric techniques are playing a critical role in understanding the relationship between the genome and the proteome. Other challenges still lie ahead. Structural determinations of proteins have now become almost routine. However, such achievements alone are not sufficient to understand protein function. Further instrumental and biochemical developments are under way that aim to enable mass spectrometry to answer these outstanding biological questions.

Suggested Reading

M. Barber et al. "Fast Atom Bombardment of Solids (FAB): A New Ion Source for Mass Spectrometry." *Journal of the Chemical Society, Chemical Communications* 7 (1981), 325–327.

K. Biemann. "A Renaissance Man in Mass Spectrometry." *Chemical and Engineering News* 48 (13 July 1970), 50–54.

R. M. Caprioli, ed. *Continuous Flow Fast Atom Bombardment Mass Spectrometry.* New York: John Wiley and Sons, 1990.

R. B. Cole, ed. *Electrospray Ionization Mass Spectrometry: Fundamentals, Instrumentation, and Applications.* New York: Wiley-Interscience, 1997.

M. Dole et al. "Molecular Beams of Macroions." *Journal of Chemical Physics* 49 (1968), 2240.

J. Duchesne et al. "The Study of Diabetes Using Naturally Enriched 13-C-Glucose." In *Proceedings of the Second International Conference on Stable Isotopes*, 282–286. Chicago: Argonne National Laboratory, U.S. Department of Energy, 1976.

MASS SPECTROMETRY	1948		1949
	Harland Wood investigates the synthesis of liver glycogen in rats using isotopically labeled compounds.	Preparative scale mass spectrometry produces 6.5 milligrams of zinc isotopes in 14 hours.	Researchers at the National Bureau of Standards describe ion cyclotron resonance as a precise means to determine the value of the Faraday constant.

HISTORY	1948		1949
	The state of Israel is founded.	Mohandas Gandhi is assassinated for his calls for peace between India's Hindus and Muslims.	The Soviet Union explodes its first atomic bomb.

J. B. Fenn et al. "Electrospray Ionization for Mass Spectrometry of Large Biomolecules." *Science* 246 (1989), 64–71.

D. A. Goldthwait; R. W. Hanson. "Harland Goff Wood." In *Biographical Memoirs of the National Academy of Sciences, 1877–.* Vol. 69, 395–428. Washington, D.C.: National Academy Press, 1996.

A. J. Ihde. "Biochemistry II: The Dynamic Period." In *The Development of Modern Chemistry,* 643–670. New York: Dover Publications, 1984.

E. Jellum; O. Stokke; L. Eldjarn. "Application of Gas Chromatography, Mass Spectrometry, and Computer Methods on Clinical Biochemistry." *Analytical Chemistry* 45 (1973), 1099–1106.

M. Karas; F. Hillenkamp. "Laser Desorption Ionization of Proteins with Molecular Masses Exceeding 10,000 Daltons." *Analytical Chemistry* 60 (1988), 2299–2301.

P. Klein et al. "Noninvasive Detection of *Helicobacter pylori* Infection in Clinical Practice: The ^{13}C Urea Breath Test." *American Journal of Gastroenterology* 91 (1996), 690–694.

B. S. Larsen; C. N. McEwen, eds. *Mass Spectrometry of Biological Materials, Second Edition, Revised and Expanded.* New York: Dekker, 1998.

R. D. Macfarlane; D. F. Torgerson. "Californium-252 Plasma Desorption Mass Spectroscopy." *Science* 191 (1976), 920–925.

G. A. Parker. "George de Hevesy." In *Nobel Laureates in Chemistry,* edited by L. K. James, 266–271. Washington, D.C.: American Chemical Society; Philadelphia: Chemical Heritage Foundation, 1993.

D. N. Perkins et al. "Probability-Based Protein Identification by Searching Sequence Databases Using Mass Spectrometry Data." *Electrophoresis* 20 (1999), 3551–3567.

D. Rittenberg et al. "Studies in Protein Metabolism II: The Determination of Nitrogen Isotopes in Organic Compounds." *Journal of Biological Chemistry* 127 (1939), 291–299.

R. Ryhage. "Use of a Mass Spectrometer as a Detector and Analyzer for Effluents Emerging from High Temperature Gas Liquid Chromatography Columns." *Analytical Chemistry* 36 (1964), 759–764.

J. Schneider et al. "Breath Analysis of ^{13}CO$_2$ Following N-Demethylation of ^{13}C- Aminopyrine: A Measure of Liver Microsomal Function." In *Proceedings of the Second International Conference on Stable Isotopes,* 259–264. Chicago: Argonne National Laboratory, U.S. Department of Energy, 1976.

D. Schoeller; L. G. Bandini; W. H. Dietz. "Inaccuracies in Self-Reported Intake Identification by Comparison with the Doubly Labeled Water Method." *Canadian Journal of Physiology and Pharmacology* 68 (1990), 941–949.

R. Schoenheimer; D. Rittenberg. "Studies in Protein Metabolism I: General Considerations in the Application of Isotopes to the Study of Protein Metabolism. The Normal Abundance of Nitrogen Isotopes in Amino Acids." *Journal of Biological Chemistry* 127 (1939), 285–290.

B. A. Thomson; J. V. Iribarne. "Field Induced Ion Evaporation from Liquid Surfaces at Atmospheric Pressure." *Journal of Chemical Physics* 71 (1979), 4451.

H. Urey; G. Teal. "The Hydrogen Isotope of Atomic Weight Two." *Reviews of Modern Physics* 7 (1935), 34–94.

J. Watkins et al. "^{13}C-Trioctanoin: A Sensitive, Safe Test for Fat Malabsorption." In *Proceedings of the Second International Conference on Stable Isotopes,* 274–281. Chicago: Argonne National Laboratory, U.S. Department of Energy, 1976.

Direct Quotations

Pages 59–60, Schoenheimer and Rittenberg, 1939, p. 285.

Page 60, Rittenberg et al., 1939, p. 291.

Page 62, Jellum, Stokke, and Eldjarn, 1973, p. 1099.

China's long civil war ends in a Communist victory and the founding of the People's Republic of China.

Linus Pauling determines that a genetic hemoglobin abnormality is the cause of sickle-cell anemia.

The North Atlantic Treaty Organization (NATO) is founded.

Chapter **6**

Better Molecules for
Better Living

Right: The chemical structure for quinine, an important antimalarial drug derived from the bark of the cinchona plant, was determined in the mid-1930s, nearly three hundred years after its first use as a medicinal. Since the mid 1950s mass spectrometry has been an important tool for determining the chemical structure of both natural and synthetic products, thus speeding the development of new drugs.

Left: Children receiving vaccine inoculations.

Although the first synthesized drugs—aspirin and antipyrine— appeared in the 1890s, the technologies used to discover and produce medicines remained crude until the middle third of the twentieth century. Drug discovery was by and large a trial-and-error process, grounded in the use of alcohol- or water-based extraction of herbs or animal products to produce tinctures, poultices, soups, and teas. Although these pharmaceuticals derived from natural products were reasonably effective for simple ailments, they did not prove effective in fighting such infectious diseases as tuberculosis, cholera, and diphtheria, which continued to devastate the human populations worldwide. Introduction of the germ theory of disease and landmark developments in analytical instrumentation began to provide a scientific rationale on which to base disease identification and the choice of pharmaceuticals for treatment.

During the 1940s these advances were bolstered by major breakthroughs in our understanding of the immune system and the development of antibiotics. The 1950s witnessed the first appearance of protein sequencing and synthesis, tissue culture analysis, the introduction of cortisone and a vaccine to cure polio, and the discovery of interferons. By 1970 pharmaceutical manufacturers were turning out the first immunosuppressants for organ transplants and drugs to regulate blood pressure.

MASS SPECTROMETRY | **1950**

The first high-precision gas isotope ratio mass spectrometer is put into operation at the University of Chicago.

Consolidated Engineering Corporation (CEC) introduces its Model 21-103 mass spectrometer. Heated inlet systems for gas-liquid sample introduction are subsequently developed for this instrument.

HISTORY | **1950**

The Korean War begins.

Senator Joseph McCarthy begins his anti-Communist crusade.

China occupies Tibet.

Today, pharmaceutical research is a high-technology endeavor, and mass spectrometry plays a crucial role in this process. Gone are the days of trial-and-error methods of drug discovery.

The quickened pace at which medical breakthroughs were achieved during the second half of the twentieth century can be attributed in large part to the availability of new analytical techniques. Advances in materials and instrumentation facilitated the isolation of pharmacologically active compounds and characterization of their chemical and physical properties. Chromatography provided cleaner and purer extracts, while mass spectrometry, along with nuclear magnetic resonance and infrared spectroscopy, provided insights into molecular weight and structure. Gas chromatography mass spectrometry (GC-MS), the first of many so-called hyphenated mass spectrometry techniques, was especially important in this regard because it permitted pharmaceutical scientists to perform accurate qualitative and quantitative analysis of new compounds in a matter of hours.

The post–World War II period has been demonstrably the most active and productive period of pharmaceutical research to date. Human life expectancy has increased, while mortality rates attributable to disease have been lowered. The quality of life has improved as effective treatments have been developed for such nonacute conditions as osteoporosis, obesity, impotence, and male pattern baldness. Driven at least in part by the needs of this growing industry, advances in the design and performance of mass spectrometers have extended their specificity, selectivity, and sensitivity. Mass spectrometry has thus remained at the forefront of analytical techniques that have contributed to the success of the pharmaceutical industry's efforts. The continued strength of the connection between mass spectrometry and drug research is shown by the fact that more than one fifth of all presentations at the American Society for Mass Spectrometry's Annual Conference on Mass Spectrometry and Allied Topics in 2000 were devoted to pharmaceutical applications. Thus, we turn our attention to the history of this fruitful area of research.

The Modern Drug Discovery and Development Process

The discovery and development of a drug can perhaps best be described as a "funnel process" involving many interrelated steps. Thousands of potential drug candidates must be screened in order to identify a compound or new chemical entity (NCE) that has a chance of clearing drug development hurdles successfully. Once discovered, the NCE must undergo many tests before it can be passed along to the Food and Drug Administration (FDA) for approval.

MASS SPECTROMETRY	**1950**
	An improved isotope ratio mass spectrometer permits measurement of ^{13}C and ^{18}O abundance to plus or minus 0.01 percent.

HISTORY	**1950**	**1951**	
	Charles Schulz begins writing his comic strip *Peanuts*.	The first oral contraceptive is developed.	Libya gains independence from Italy.

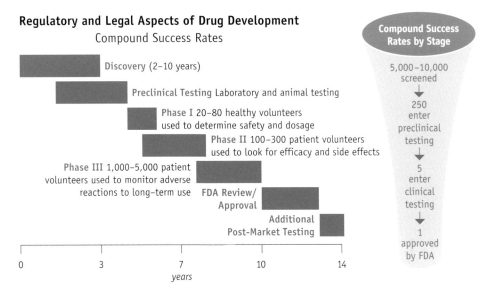

Figure 1. The drug development process. Reproduced with permission. Pharmaceutical Research and Manufacturers of America, 2001 Industry Profile, PhRMA, Washington, D.C., 2001.

Regulatory and Legal Aspects of Drug Development
Compound Success Rates

Discovery (2–10 years)

Preclinical Testing Laboratory and animal testing

Phase I 20–80 healthy volunteers used to determine safety and dosage

Phase II 100–300 patient volunteers used to look for efficacy and side effects

Phase III 1,000–5,000 patient volunteers used to monitor adverse reactions to long-term use

FDA Review/Approval

Additional Post-Market Testing

years

0   3   7   10   14

Compound Success Rates by Stage

5,000–10,000 screened
↓
250 enter preclinical testing
↓
5 enter clinical testing
↓
1 approved by FDA

Source: PhRMA, based on data from the Center for the Study of Drug Development, Tufts University, 1995.

Figure 1 describes the current drug development process. The cost of bringing one NCE to market can be as high as $300 million, and the process can take over ten years.

Early activities in this multistep process include identification of specific disease targets and generation of lead therapeutic agents by isolation from natural sources or by synthesis. Potency must be determined and dosage levels developed. With a potential new therapeutic agent in hand, safety and efficacy must be established, first through toxicological evaluation and then through studies in humans. Pharmacokinetic and pharmacodynamic evaluations provide the basis for the establishment of dosage regimens for normal as well as special populations. Biotransformation studies clarify ways in which the drug is processed in the body and elucidate potentially harmful metabolic pathways. Parallel to the clinical activities is a massive effort in manufacturing to scale up the synthetic route from the benchtop to a time- and cost-efficient large-scale industrial process.

The Early Days of Mass Spectrometry in Pharmaceutical Analysis

For more than forty years mass spectrometry has been used to study small molecules (up to 600 daltons) of pharmacological interest. In the 1960s mass spectrometric contributions in support of drug discovery consisted primarily of molecular weight determinations and structure proofs of synthetic routes for natural products. Magnetic-sector mass spectrometers fitted with electron ionization sources were commonly used. By current standards these instruments were expensive, cumbersome, and difficult to operate. Nevertheless synthetic chemists in the pharmaceutical industry found the instruments critical to their analytical needs.

1952

A. O. Nier and E. G. Johnson design a double-focusing mass spectrometer that focuses ions onto a single point.

A. J. Martin and R. L. Synge receive the Nobel Prize in chemistry for their invention of partition chromatography.

CEC introduces SpectroSADIC, an analog-to-digital converter for the acquisition of up to forty selected peaks in the mass range of 12 to 150 daltons.

1952

Gamal Abdel Nasser comes to power in Egypt after King Farouk I is overthrown in a coup.

Jonas Salk introduces a polio vaccine.

New ionization techniques that were developed in the 1970s and 1980s enabled mass spectral characterization and study of new classes of compounds. With chemical ionization (CI), molecular weights could be determined even for chemically fragile molecules. Field ionization (FI), field desorption (FD), and fast atom bombardment (FAB) enabled non-volatile and polar molecules to be studied by mass spectrometric techniques. The coupling of gas chromatography to mass spectrometry provided on-line identification of the constituents of complex mixtures. Early steps in peptide sequencing were accomplished by GC-MS analysis of extracts obtained from Edman degradation of intact proteins.

Using electron ionization, chemical ionization, and field ionization, several metabolism-related issues critical to drug development were addressed by pharmaceutical researchers. Early in the 1970s GC-MS played an important role in the identification of N-oxide metabolites and provided insights into the possible involvement of N-oxides in cancer. However, because of its limited thermal stability, volatility, and polarity, amine-N-oxide formation was not detected by GC-MS. Studies also showed the possibility of N-oxide metabolite decomposition when plasma samples are alkalized during extraction using sodium hydroxide rather than sodium carbonate. FI-MS was used to study the metabolism and disposition of diamino piperidino-pyrimidine (a drug currently used to treat male hair loss) while it was being developed as a drug to combat hypotension. FI-MS studies showed that a lesion on the right atrium of dogs that were administered the drug was not associated with any of the drug's metabolites. Furthermore, FI-MS studies established that monkeys and humans possess similar metabolism and disposition patterns for this drug. These studies were important for establishing the drug's safety.

The commercialization of quadrupole mass spectrometers in the 1970s provided mass selective devices that could be coupled to gas chromatographs more easily and reliably than their magnetic sector counterparts. Qualitative and quantitative analysis of many small-molecule pharmaceuticals was accomplished with the electron and chemical ionization sources that were standard equipment on these quadrupole instruments. Chemical modification (derivatization) of polar substituents (e.g., -OH, -COOH, -NH) on these small molecules improved their thermal stability and increased their volatility, rendering them suitable for mass spectrometric analysis. The simplicity of coupling three quadrupole instruments to one another in series—two for mass analysis and one in between for collision-induced dissociation (CID)—put the means for additional selectivity in quantitative analysis at the disposal of chemists engaged in all aspects of pharmaceutical research. With triple quadrupole mass spectrometry, the use of low-energy collisions produced fragmentation patterns rich in structural information, thus facilitating the identification of unknowns for the elucidation of metabolic pathways. A recent example of this is the case of the oral antidiabetic agent triazolidinedione (troglitazone). Tandem mass spectrometry was used to characterize some of the possible reactive metabolites associated with drug-induced toxicity.

MASS SPECTROMETRY	**1952**		
	The absolute rate theory is developed for mass spectra of polyatomic molecules.	Victor Talrose discovers that elementary ion-molecule reactions of organic compounds have no activation barrier.	The American Society for Testing and Materials (ASTM) Committee E-14 on Mass Spectrometry is organized.
HISTORY	**1952**		
	The first hydrogen bomb is exploded by the United States at Bikini Atoll.	The first commercial jet airliner, the DeHavilland Comet, enters service.	

Tandem Mass Spectrometry

The earliest tandem mass spectrometers (MS-MS), developed in the mid-1960s, were actually designed to perform fundamental ion-molecule reaction studies. The first mass spectrometer (MS1) was used to select an ion to collide with a neutral molecule in a gas cell between the two instruments. The ionic products from that collision were then analyzed with the second mass spectrometer (MS2).

While those interested in fundamental studies pursued their research using the MS-MS, some scientists saw the tandem instrument as more of an analytical tool, along the lines of GC-MS. In the case of the MS-MS, the GC is replaced by MS1. The collision cell between the two instruments is still necessary to impart energy to, and thus promote fragmentation of, the ion selected by MS1. Otherwise, the spectrum obtained by MS2 would not provide much useful information about the structure of the selected ion.

The earliest tandem mass spectrometry experiments were performed in the 1970s using the electrostatic sector of a double-focusing mass spectrometer to scan the energies of ions emerging from the collision cell. The resulting low-resolution spectra provided fundamental information about the kinetic energy released in the ion-molecule collision. Subsequent instrumental improvements led to the development of a variety of instrument combinations to perform tandem mass spectrometry experiments. Examples include linked sector, triple quadrupole, and quadrupole–time-of-flight mass spectrometers. Tandem mass spectrometry is an indispensable tool in proteomic applications today.

Reference
J. H Futrell and C. D. Miller. "Tandem Mass Spectrometer for Study of Ion-Molecule Reactions." *Review of Scientific Instruments* 37 (1966), 1521–1526.

GC-MS, with various ionization methods, has been used to search out and identify potential drugs. It has rapidly and specifically identified and differentiated more than ten alkylamine antihistamines. Screening was later extended to more than forty antihistamines through the combined use of electron and chemical ionization methods. GC-MS also played an important role in bringing an antiviral drug, α-methyl tricyclodecane methanamine, to market. The work on this drug showed that mass spectrometry was successful in bioanalysis because of its high specificity, sensitivity, and throughput.

A life-threatening situation would not have been recognized without the use of mass spectrometric techniques in metabolic studies of a drug composed of butanedioic acid, N,N-dimethyl-phenyl-pyridinyl-ethoxy-ethanamine, and hydroxyl-methyl-pyridinedimethanol-hydrochloride. Studies showed the presence of several phenolic metabolites in rat and monkey excreta that may have resulted from electrophilic arene oxide intermediates in the drug's metabolic pathway. These intermediates are capable of reacting with cellular nucleophiles, thereby causing significant cellular damage. Identification of metabolites in this work involved extensive sample extraction and modification. Although now overshadowed by liquid chromatography mass spectrometry (LC-MS) in pharmaceutical applications, GC-MS continues to play a key role in particular areas of pharmaceutical analysis, as demonstrated in

1953

Field ionization phenomena are observed at the University of Chicago.

A vacuum lock for solids analysis is first described at Argonne National Laboratory.

Wolfgang Paul publishes early papers on quadrupole mass spectrometers and ion-trap detectors.

1953

James Watson and Francis Crick, working with the help of data obtained by Rosalind Franklin and Maurice Wilkins, determine the double-helical structure of DNA.

Activist Malcolm Little changes his name to Malcolm X.

Julius and Ethel Rosenberg are executed for giving atomic bomb secrets to the Soviets.

the recent biotransformation studies of dimethoxy-phenyl-methyl-amino-hydroxy-propoxy-quinolinone, a novel inotropic drug that affects the strength of the heart's contraction, which is currently under clinical development for the treatment of congestive heart failure.

The advent of fast atom bombardment in the early 1980s provided the pharmaceutical industry with a tool that could desorb ionic, thermally labile, and low-volatility drug molecules and metabolites directly from the solution phase into the gas phase. The sensitivity of the technique could be enhanced by using chemical derivatization. The use of static FAB along with flow-FAB and tandem mass spectrometry allowed pharmaceutical scientists to study natural products that occur at very low concentrations and to characterize modifications of them in such important biochemical processes as deamidation, oxidation, glycosylation, phosphorylation, and disulfide bond formation. FAB, in combination with tandem mass spectrometry, has also been successful in the quantitative analysis of benzazepine, a potent and selective antagonist of the alpha-2 adenoreceptor. In 1982 the critical observation that mass spectral features common to both drugs and their metabolites could be used for rapid detection and characterization of metabolites was made. FAB tandem MS or MS-MS was used in the detection of glucuronide and sulfate conjugates. In 1993 FAB, in combination with tandem mass spectrometry, was used to show for the first time that the oxygen of pyridine-N-oxide could undergo glucuronidation.

Another area that has been advanced by FAB is the investigation of reactive drug metabolites. Covalent modification of such macromolecules as DNA have been associated with cellular changes that result in tumor growth and cell death, thus making the identification of reactive metabolites an important issue during drug discovery and development. FAB-MS and FAB-MS-MS have also been used to show that N-acetyl-S-(N-methyl-carbamoyl) cysteine is chemically reactive under conditions of physiological pH and temperature. Similarly, FAB-MS and MS-MS have been used to study seven antihistamines, their N-oxides, and related conjugated metabolites. Using FAB ionization, reduction of N-oxide metabolites into the original parent drug has been observed. This conversion has been shown to be a matrix-dependent solution-phase process.

The separation of the analyte from such complex sample matrices as plasma or urine before ionization is critical to achieving sensitivity and specificity in the quantitative analysis of pharmaceutical applications. For this reason pharmaceutical applications involving quantitative analysis by electron ionization (EI)-solids probe or FAB have been limited. Instead, when sensitivity or specificity, or both, have been identified as especially important in quantitative analysis, GC-MS and GC-MS-MS methods have been used to analyze the components of drugs and related metabolites in plasma and urine. Modifying polar substituents with perfluorinated reagents has enhanced the ability of certain drug analytes to capture electrons, making negative chemical ionization GC-MS the most sensitive analytical technique available for quantitative analysis of these species. Other analytical methods for numerous drugs and their metabolites have been published as well. Examples include

MASS SPECTROMETRY

1953

The first annual Conference on Mass Spectrometry and Allied Topics is held solely under the aegis of the ASTM Committee E-14.

1954

High-resolution mass spectrometry is developed at Imperial Chemical Industries in the United Kingdom for the study of organic compounds.

HISTORY

1953

Soviet dictator Joseph Stalin and Soviet composer Sergei Prokofiev die on the same day, 6 March.

Edmund Hillary and Tenzing Norgay become the first people to climb Mt. Everest.

1954

The U.S. Supreme Court rules that racial segregation is unconstitutional in the case *Brown v. Board of Education of Topeka*.

angiotensin-converting enzyme inhibitors and their active desethyl metabolite, selective dopaminergic drugs, and the statins.

The 1990s: Coming of Age for LC-MS in Pharmaceutical Analysis

High-performance liquid chromatography was coupled with ultraviolet detection (HPLC-UV) in the 1970s. It quickly found widespread use as an analytical tool in the pharmaceutical industry. For example, polar and thermally labile drug molecules were much more amenable to separation by liquid chromatographic methods than by gas chromatographic methods, and LC worked well with UV detectors. Of course, HPLC-UV lacked the selectivity and specificity of mass spectrometry, but coupling a liquid chromatograph to a mass spectrometer proved exceedingly difficult. At the time, no appropriate mechanism was available to couple the buffered liquid effluent from the HPLC (often high in aqueous content) to the vacuum of the mass spectrometer. During the next twenty years approximately twenty-five different interfaces were developed to couple LC with MS. Thermospray (TS) and particle-beam (PB) ionization-interface techniques, for example, showed early promise in pharmaceutical applications. PB-MS provided high-efficiency separation as well as mass spectra with high information content. Its applicability was limited, however, when low concentrations of sample materials were used.

Thermospray was a soft ionization technique that could overcome many of the limitations of electron and chemical ionization. For example, it could function at liquid chromatographic flow rates in the range of 0.5 to 2.0 milliliters per minute with mobile phases containing large percentages of water and buffer. As a result TS made coupling of HPLC to MS possible. By the mid-1980s LC-TS-MS was used in the pharmaceutical industry for organic synthesis confirmation and qualitative and quantitative studies. Although sample modification by derivatization was no longer necessary in cases in which TS was used, it did increase the detection limits of mass spectrometers by several orders of magnitude.

While qualitative metabolite identification studies using electron, chemical, and field ionization techniques have been performed, the introduction of TS enhanced the role of mass spectrometry in metabolism-related studies. Major and minor urinary metabolites of an antiepileptic drug, dichlorophenyl triazine diamine, which had never been characterized before using conventional HPLC and radioactive tracer detection methods, were characterized by LC-TS-MS using samples containing only billionths of a gram of sample material. In an interesting comparison study, metabolite identification data for dibromo-hydroxyphenyl-ethyl-benzofuranyl-methanone—obtained using LC-TS-MS—were compared with data from EI-MS. Obtaining information-rich EI spectra required relatively large amounts of material isolated by semipreparative methods, whereas LC-TS-MS provided sufficient structural information with smaller amounts of sample material that did not require semipreparative cleanup. Furthermore, LC-TS-MS provided information on minor metabolites as well as some phase II metabolites that were not detectable when EI-MS was

Roger Bannister runs a mile in less than four minutes.

Joseph E. Murray of Harvard carries out the first successful human kidney transplant.

The Eisenhower administration suspends the security clearance of respected physicist J. Robert Oppenheimer.

Better Molecules for Better Living

Alphabet Soup Mass Spectrometry

The wide variety of ionization methods and mass analyzers that have been developed over the years and the combination of instruments that separate mixtures have led to a bewildering proliferation of technical terms, abbreviations, and initialisms. Every field has its jargon, but mass spectrometry has spawned a strikingly unique language of its own.

Probably the earliest assignment of initials began with descriptions of different methods used to ionize molecules of varying sizes: EI for electron ionization, CI for chemical ionization, FAB for fast atom bombardment. The list goes on. When gas chromatography mass spectrometry was introduced, more abbreviations appeared. GC-MS was followed by LC-MS for liquid chromatography mass spectrometry, and CE-MS or CZE-MS, for capillary electrophoresis separations. Tandem mass spectrometry added a whole new dimension to the alphabet soup lexicon. Combinations include EBE, BEB, BEBE, BEQ, and Q-TOF, to mention a few. Some of the more important abbreviations and their full descriptions are listed below.

APCI	Atmospheric pressure chemical ionization		**FD**	Field desorption
B	Magnetic sector (when used with E or Q, or both)		**FI**	Field ionization
			FT	Fourier transform
CAD	Collision-activated dissociation		**GC**	Gas chromatograph
CE	Capillary electrophoresis		**ICR**	Ion-cyclotron resonance
CI	Chemical ionization		**IR**	Isotope ratio
CID	Collision-induced dissociation		**IS**	Ionspray
DS	Data system		**LC**	Liquid chromatograph
E	Electrostatic sector (when used with B or Q, or both)		**MALDI**	Matrix-assisted laser desorption ionization
EI	Electron ionization		**Q**	Quadrupole mass filter
ESI	Electrospray ionization		**TOF**	Time-of-flight
FAB	Fast atom bombardment		**TS**	Thermospray

used. Although detection of phase II metabolites can be difficult owing to thermal degradation of the sample, detection and characterization of glucuronide and sulfate metabolites have been achieved.

Although thermospray ionization facilitated the application of LC-MS to pharmaceutical problems, electrospray ionization (ESI), ionspray (IS), and atmospheric pressure chemical ionization (APCI) represented a new and more advanced generation of ionization-interfacing technologies. The successful application of these new ionization techniques depended on a critical change in the coupling technology in LC-MS instruments. The liquid effluent from the liquid chromatograph was converted to a gas by supersonic expansion through a nozzle. Further refinements of the technique included enrichment of analyte composition in the stream by evaporation of the solvent through the use of counter-current gases and the substitution of volatile buffers in place of the traditional phosphates.

MASS SPECTROMETRY	1955	1956
	Consolidated Engineering Corporation changes its name to Consolidated Electrodynamics Corporation.	Proton affinity determinations are made by using mass spectrometry.

HISTORY	1954	1956
	Rosa Parks is arrested for refusing to give up her seat to a white passenger on a city bus in Montgomery, Alabama.	Egypt nationalizes the Suez Canal, leading to war with Israel, France, and Britain.

Initially, electrospray ionization was used in the study of large biomolecules, such as proteins. More recently, especially during the 1990s, LC-ESI-MS and LC-APCI-MS replaced HPLC-UV in pharmaceutical research. Using these new ionization interfaces, LC-MS-MS has been widely employed for impurity analyses and structure elucidation of drug candidates. Moreover, sensitivity is more important for quantitative analysis as more potent drugs have been developed and time-released drugs have been used for drug delivery. Consider, for example, the distribution of quantitative assays among other analytical techniques used in the pharmaceutical research laboratory. In 1990 roughly 85 percent of drug analyses were performed using HPLC, 12 percent using GC-MS, and 3 percent using LC-MS. Ten years later the numbers had switched completely. More than 80 percent of quantitative assays had been performed using LC-MS, nearly 15 percent using HPLC, and the remaining few percent using GC-MS.

Coupling HPLC to a mass spectrometer improved sensitivity and specificity and also brought about dramatic changes in the speed of analysis. This improved performance matched changes occurring in the drug discovery process itself. Sample separations that used to take up to forty-five minutes could be routinely completed in less than ten minutes. Consequently, mass spectrometry became an important tool in natural-product dereplication, and it was also used to determine the quality and confirmation of compound identities in combinatorial libraries. Today, data on molecular weights of potential drug candidates can be made available to synthetic chemists in a matter of minutes. Also speeding up the process is the use of LC-MS and LC-MS-MS in accelerated test or forced degradation studies to characterize chemical degradants in pharmaceuticals.

The speed with which analyses can be completed has promoted an interest in increased sample throughput, particularly for quantitative analysis of pharmaceuticals in biological fluids. Shorter and narrower columns, in combination with higher HPLC flow rates, are used to achieve higher sample throughput for chromatography. Moreover, the introduction of four- and eight-channel multiplexed ESI interfaces promises further increases in sample throughput for mass spectrometry. A four-column chromatography system capable of operating under isocratic as well as gradient conditions has recently been introduced. Paralleling these developments in analytical instrumentation are the processes of manipulating the sample materials. The pharmaceutical industry has invested significant resources in this field. Sample purification, on-line sample processing, semi-automated ninety-six-well extraction, and sample collection and handling systems have been developed to work with automated LC-MS-MS systems.

The Future: Old Techniques with New Promise, New Techniques with More Promise

Mass spectrometry can be expected to play a critical role in the analysis of pharmaceutical compounds in the twenty-first century. In the race to obtain more and better information on pharmaceutical compounds early in their life cycles, current applications of mass spectrometry

Fred McLafferty proposes a mechanism for γ-hydrogen transfer that becomes known as the McLafferty rearrangement.

Steroids are first analyzed using mass spectrometry.

1957

Amoco researcher Seymour Myerson confirms the existence of the tropylium ion.

Democratic reforms in Hungary prompt a Soviet invasion.

The United States sends military advisers to South Vietnam.

1957

The Soviet satellite *Sputnik I* becomes the first human-made object to orbit Earth.

in the drug discovery process fall into four general categories: analytical specificity by high-resolution tandem mass spectrometry systems, new high-throughput screening methods to obtain metabolic data, studies of drug transport, and investigations of protein binding.

The identification of active drug target components, isolated from the plethora of sources of natural products and target candidates, has been aided by the recent introduction of several hybrid, time-of-flight, ion-trap, and triple-quadrupole mass spectrometers with high resolving powers. These systems have brought high resolution within easy reach of pharmaceutical researchers. Automated Edman degradation, the standard method of choice for peptide analysis, is quickly being replaced by mass spectrometry because of its inherent sensitivity, its ability to detect post-translational modifications, and the possibility of rapidly analyzing multi-component mixtures that cannot otherwise be separated chromatographically. Nanoelectrospray ionization has dramatically reduced sample consumption, making it possible to identify and characterize extremely small quantities of biomolecular species.

Pharmaceutical researchers are especially interested in obtaining metabolic information during the drug discovery process. With this information they hope to minimize compound failure in the drug discovery stage and also cope with large numbers of lead candidates introduced using combinatorial libraries. Pharmaceutical scientists must choose lead compounds with minimal metabolic liability, and they must also try to mitigate the possibility of drug-drug interactions. High-throughput screening methods based on a ninety-six-well plate format have been developed to investigate microsomal stability, metabolism, and the potential for drug-drug interactions.

Liquid chromatography mass spectrometry has also enabled pharmaceutical researchers to study drug transport processes. In order to predict the in vivo bioavailability of a drug, transport of the drug across the human colon adenocarcinoma cell line (Caco-2) is being used as an absorption model. High-throughput LC-MS-MS methods have expedited in an unprecedented way the availability of transport information on new drugs.

The degree of protein binding is also an important screening parameter in the drug discovery process. Traditional methods for studying protein binding include equilibrium dialysis, liquid chromatography with size-exclusion techniques, and LC-UV with the bovine albumin column. Introduction of ESI-MS has offered a sensitive and selective method enabling protein-binding studies to be conducted using high-throughput methods.

Advances in genomics, proteomics, and related technologies, including mass spectrometry, have led to a wealth of potential drug targets. Pharmaceutical researchers will continue to rely on new ionization techniques, especially matrix-assisted laser desorption ionization. Quantitative studies using this versatile ionization technique for desorption and ionization are also beginning to appear.

Enhancements in sensitivity now seem likely with accelerator mass spectrometry (AMS). Although the operation of AMS differs significantly from conventional mass spectrometers and gives no molecular weight or structural information, the technique is sensitive

MASS SPECTROMETRY	1957	
	Trimethylsilyl esters are used as derivatives for mass spectrometry.	GC-MS is first demonstrated at the Phillip-Morris Company.

HISTORY	1957		
	President Dwight Eisenhower sends federal troops to Little Rock, Arkansas, to enforce public school desegregation.	Ghana gains independence from Britain.	François "Papa Doc" Duvalier comes to power in Haiti.

enough to detect radioisotopes present at sample concentrations corresponding to just a handful of atoms. Pharmaceutical companies have been somewhat slow to embrace AMS technology because of the size and expense of the instrumentation and the necessity for radiolabeling the administered drugs. The technique holds great promise, however, for pharmacokinetic studies and elucidation of metabolic pathways with low-dose ^{14}C-labeled drugs in plasma. Many pharmaceutical companies are making use of commercial facilities that offer AMS technology.

The pharmaceutical industry has embraced mass spectrometry because of its ability to reduce the time horizon for drug development. In addition to producing more effective drugs in less time, this high-precision analytical technique has also enabled the pharmaceutical industry to maintain high-quality standards for its products over the long term. By 2005 pharmaceutical patents worth more than $35 billion will expire. Corporate mergers, fragmenting global markets, and political and policy initiatives will also put pressure on firms to develop more aggressive innovative procedures.

In the meantime mass spectrometry will continue to be an active participant in the process, and the industry is already beginning to see the effects of major improvements in instrumentation. High-throughput sample analysis, in which two thousand samples can be run in just one day, is already a reality. Fourier transform ion cyclotron resonance mass spectrometry in conjunction with ESI is being applied as a high-throughput ligand-binding screening tool by the biotechnology and pharmaceutical industries. Similarly in automation, intelligent decision-making software is being developed to improve the operation of laboratory mass spectrometers with minimum user interactions. Several major mass spectrometer manufacturers are developing metabolite identification software that is capable of providing accurate mass data and correlation comparison values of metabolites with respect to a specific parent drug. Undoubtedly, mass spectrometry has started to play a major role in proteomics, which in turn is expected to give new insights into disease mechanisms and improve drug discovery strategies to produce novel therapeutics.

Suggested Reading

B. L. Ackermann et al. "Rapid Analysis of Antibiotic-Containing Mixtures from Fermentation Broths by Using Liquid Chromatography–Electrospray Ionization–Mass Spectrometry and Matrix-Assisted Laser Desorption Ionization-Time-of-Flight-Mass Spectrometry." *Journal of the American Society for Mass Spectrometry* 7 (1996), 1227–1237.

M. C. Allen; T. S. Shah; W. W. Day. "Rapid Determination of Oral Pharmacokinetics and Plasma Free Fraction Using Cocktail Approaches: Methods and Application." *Pharmaceutical Research* 15 (1998), 93–97.

R. S. Annan; R. W. Giese; P. Vouros. "Detection and Structural Characterization of Amino Polyaromatic Hydrocarbon-Deoxynucleoside Adducts Using Fast Atom Bombardment and Tandem Mass Spectrometry." *Analytical Biochemistry* 191 (1990), 86–95.

1958			1959
CEC introduces Mascot, the first commercial mass spectrum digitizer.	The Bendix time-of-flight mass spectrometer is introduced, and multiple ion monitoring is demonstrated.	The first mass spectrometer measurements of the Earth's atmosphere are made by quadrupole mass filter at the U.S. Naval Research Laboratory.	Mass spectrometry is used at MIT for peptide and oligonucleotide sequencing.

1958		1959	
Egypt and Syria form the United Arab Republic.	The National Aeronautics and Space Administration (NASA) is created.	Fidel Castro comes to power in Cuba.	Louis and Mary Leakey unearth the first *Homo habilis* fossils in Olduvai Gorge, Tanzania.

A. Apffel; M. L. Perry. "Quantitation and Linearity for Particle-Beam Liquid Chromatography–Mass Spectrometry." *Journal of Chromatography* 554 (1991), 103–118.

R. E. Banks et al. "Proteomics: New Perspectives, New Biomedical Opportunities." *Lancet* 356 (2000), 1749–1756.

M. Barber et al. "Fast Atom Bombardment of Solids as an Ion Source in Mass Spectrometry." *Nature* 293 (1981), 270–275.

K. Biemann; H. A. Scoble. "Characterization by Tandem Mass Spectrometry of Structural Modifications in Proteins." *Science* 237 (1987), 992–998.

K. Biemann; W. Vetter. "Separation of Peptide Derivatives by Gas Chromatography Combined with the Mass Spectrometric Determination of the Amino Acid Sequence." *Biochemical and Biophysical Research Communications* 3 (1960), 578–584.

C. R. Blakley; M. L. Vestal. "Thermospray Interface for Liquid Chromatography/Mass Spectrometry." *Analytical Chemistry* 55 (1983), 750–754.

A. P. Bruins; T. R. Covey; J. D. Henion. "Ion Spray Interface for Combined Liquid Chromatography/ Atmospheric Pressure Ionization Mass Spectrometry." *Analytical Chemistry* 59 (1987), 2642–2646.

H. Bu et al. "High-Throughput Caco-2 Cell Permeability Screening by Cassette Dosing and Sample Pooling Approaches Using Direct Injection/On-Line Guard Cartridge Extraction/Tandem Mass Spectrometry." *Rapid Communications in Mass Spectrometry* 14 (2000), 523–528.

Z. I. Cai; A. K. Sinhababu; S. Harrelson. "Simultaneous Quantitative Cassette Analysis of Drugs and Detection of Their Metabolites by High Performance Liquid Chromatography/Ion Trap Mass Spectrometry." *Rapid Communications in Mass Spectrometry* 14 (2000), 1637–1643.

D. I. Carroll. "Atmospheric Pressure Ionization Mass Spectrometry Corona Discharge Ion Source for Use in Liquid Chromatograph–Mass Spectrometer—Computer Analytical System." *Analytical Chemistry* 47 (1975), 2369–2373.

N. J. Clarke et al. "Systematic LC/MS Metabolite Identification in Drug Discovery." *Analytical Chemistry* 73 (2001), 430A–439A.

T. N. Decaestecker et al. "Evaluation of Automated Single Mass Spectrometry to Tandem Mass Spectrometry Function Switching for Comprehensive Drug Profiling Analysis Using a Quadrupole Time-of-Flight Mass Spectrometer." *Rapid Communications in Mass Spectrometry* 14 (2000), 1787–1792.

M. V. Doig; R. A. Clare. "Use of Thermospray Liquid Chromatography–Mass Spectrometry to Aid in the Identification of Urinary Metabolites of a Novel Antiepileptic Drug, Lamotrigine." *Journal of Chromatography* 554 (1991), 181–189.

J. G. Dorsey et al. "Liquid Chromatography: Theory and Methodology." *Analytical Chemistry* 70 (1998), 591R–644R.

C. Enjalbal; J. Martinez; J.-L. Aubagnac. "Mass Spectrometry in Combinatorial Chemistry." *Mass Spectrometry Reviews* 19 (2000), 139–161.

J. Ermer; M. Vogel. "Applications of Hyphenated LC-MS Techniques in Pharmaceutical Analysis." *Biomedical Chromatography* 14 (2000), 373–383.

MASS SPECTROMETRY   **1959**

A gas chromatograph is interfaced to a time-of-flight mass spectrometer at Dow Chemical.

HISTORY   **1959**

The Soviet space probe *Lunik 2* is the first craft to land on the Moon.

The first xerographic photocopiers are sold commercially.

The International Geophysical Year begins.

E. K. Fukuda; W. A. Garland. "Applications in Biomedical Mass Spectrometry: The Rimantadine Story." In *Biological Mass Spectrometry, Proceedings of the Second International Symposium on Mass Spectrometry and the Health and Life Sciences.* Edited by A. L. Burlingame and J. A. McCloskey, 549–566. Amsterdam: Elsevier, 1990.

D. A. Ganes; K. W. Hindmarsh; K. K. Midha. "Doxylamine Metabolism in Rat and Monkey." *Xenobiotica* 16 (1986), 781–794.

A. M. Gioacchini et al. "Electrospray Ionization, Accurate Mass Measurements and Multistage Mass Spectrometry Experiments in the Characterization of Stereoisomeric Isoquinoline Alkaloids." *Rapid Communications in Mass Spectrometry* 14 (2000), 1592–1599.

J. D. Henion; B. A. Thomson; P. H. Dawson. "Determination of Sulfur Drugs in Biological Fluids by Liquid Chromatography/Mass Spectrometry." *Analytical Chemistry* 54 (1982), 451–456.

S. H. Hoke et al. "Transformations in Pharmaceutical Research and Development, Driven by Innovations in Multidimensional Mass Spectrometry-Based Technologies." *International Journal of Mass Spectrometry* 212 (2001), 135–196.

M. Jemal. "High-Throughput Quantitative Bioanalysis by LC/MS/MS." *Biomedical Chromatography* 14 (2000), 422–429.

K. Kassahun et al. "Studies on the Metabolism of Troglitazone to Reactive Intermediates in Vitro and in Vivo. Evidence for Novel Biotransformation Pathways Involving Quinone Methide Formation and Thiazolidine-dione Ring Scission." *Chemical Research in Toxicology* 14 (2001), 62–70.

D. B. Kassel. "Combinatorial Chemistry and Mass Spectrometry in the 21st Century Drug Discovery Laboratory." *Chemical Reviews* 101 (2001), 255–267.

M. Kinter; S. Singh; R. A. Felder. "Quantitation of Selective Dopaminergic Drugs in Plasma by Gas Chromatography–Mass Spectrometry Following Solid-Phase Extraction." *Journal of Chromatography, Biomedical Applications* 496 (1989), 201–208.

M. Kitani et al. "Biotransformation of the Novel Inotropic Agent Toborinone (OPC-18790) in Rats and Dogs: Evidence for the Formation of Novel Glutathione and Two Cysteine Conjugates." *Drug Metabolism Disposition* 25 (1997), 663–674.

J. O. Lay; C. L. Holder; W. M. Cooper. "Characterization of Seven Antihistamines, Their N-oxides and Related Metabolites by Fast Atom Bombardment Mass Spectrometry and Fast Atom Bombardment Tandem Mass Spectrometry." *Biomedical and Environmental Mass Spectrometry* 18 (1989), 157–167.

M. S. Lee; E. H. Kearns. "LC/MS in Drug Development." *Mass Spectrometry Reviews* 18 (1999), 187–279.

J. A. Loo et al. "Application of Mass Spectrometry for Target Identification and Characterization." *Medicinal Research Reviews* 19 (1999), 307–319.

H. Maurer; K. Pfleger. "Identification and Differentiation of Alkylamine Antihistamines and Their Metabolites in Urine by Computerized Gas Chromatography–Mass Spectrometry." *Journal of Chromatography* 430 (1988), 31–41.

S. R. Needham; P. R. Brown. "The Role of the Column for the Analysis of Drugs and Other Components by HPLC/ESI/MS: Part 1." *American Pharmaceutical Review* 3 (2000), 45–50.

1960

Mass spectrometers analyze and monitor air quality in submarines.	Bendix Corporation introduces the first direct insertion (solids) probe at the Pittsburgh Conference.	CEC introduces its Model 21-110 high-resolution double-focusing mass spectrograph.

1960

The Belgian Congo becomes the independent nation of Zaire.	Theodore Harold Maiman develops the first laser.	The first communications and weather satellites are sent into orbit.	Americans prepare for nuclear war by constructing fallout shelters.

W. M. A. Niessen; U. R. Tjaden; J. Van der Greef. "Strategies in Developing Interfaces for Coupling Liquid Chromatography and Mass Spectrometry." *Journal of Chromatography* 554 (1991), 3–26.

E. J. Oliveira; D. G. Watson. "Liquid Chromatography–Mass Spectrometry in the Study of the Metabolism of Drugs and Other Xenobiotics." *Biomedical Chromatography* 14 (2000), 351–372.

J. M. Onorato et al. "Selected Reaction Monitoring LC-MS Determination of Idoxifene and Its Pyrrolidinone Metabolite in Human Plasma Using Robotic High-Throughput, Sequential Sample Injection." *Analytical Chemistry* 73 (2001), 119–125.

R. J. Perchalski; R. A. Yost; B. J. Wilder. "Structural Elucidation of Drug Metabolites by Triple Quadrupole Mass Spectrometry." *Analytical Chemistry* 54 (1982), 1466–1471.

Pharmaceutical Industry Profile. Washington, D.C.: Pharmaceutical Research and Manufacturers of America, 1999.

B. N. Pramanik; P. L. Bartner; G. Chen. "The Role of Mass Spectrometry in the Drug Discovery Process." *Current Opinion in Drug Discovery and Development* 2 (1999), 401–417.

A. D. Rodrigues et al. "Measurement of Liver Microsomal Cytochrome P450 (CYP2D6) Activity Using [O-methyl-14C]dextromethorphan." *Analytical Chemistry and Biochemistry* 219 (1994), 309–320.

P. Rudewicz; K. M. Straub. "Rapid Structure Elucidation of Catecholamine Conjugates with Tandem Mass Spectrometry." *Analytical Chemistry* 58 (1986), 2928–2934.

A. B. Sage; D. Little; K. Giles. "Using Parallel LC-MS and LC-MS-MS to Increase Sample Throughput." *LC-GC* 18 (2000), S20–S29.

K. Sakamoto; Y. Nakamura. "Urinary Metabolites of Pinacidil. II. Species Difference in the Metabolism of Pinacidil." *Xenobiotica* 23 (1993), 649–656.

K. A. Sannes-Lowery et al. "Fourier Transform Ion Cyclotron Resonance Mass Spectrometry as a High Throughput Affinity Screen to Identify RNA Binding Ligands." *Trends in Analytical Chemistry* 19 (2000), 481–491.

H. Shioya; M. Shimojo; Y. Kawahara. "Determination of a New Angiotensin-Converting Enzyme Inhibitor (CS-622) and Its Active Metabolite in Plasma and Urine by Gas Chromatography–Mass Spectrometry Using Negative Ion Chemical Ionization." *Journal of Chromatography, Biomedical Applications* 496 (1989), 129–135.

K. M. Straub; P. Levandoski. "Quantitative Analysis of an N-oxide Metabolite by Fast-Atom Bombardment Tandem Mass Spectrometry." *Biomedical Mass Spectrometry* 12 (1985), 338–343.

R. C. Thomas; H. J. Harpootlian. "Metabolism of Minoxidil, a New Hypotensive Agent. II. Biotransformation Following Oral Administration to Rats, Dogs, and Monkeys." *Journal of Pharmaceutical Sciences* 64 (1975), 1366–1371.

M. D. Threadgill et al. "Metabolism of N-methylformamide in Mice: Primary Kinetic Deuterium Isotope Effect and Identification of S-(N-methylcarbamoyl)glutathione as a Metabolite." *Journal of Pharmacology and Experimental Therapeutics* 242 (1987), 312–319.

K. W. Turteltaub; J. S. Vogel. "Bioanalytical Applications of Accelerator Mass Spectrometry of Pharmaceutical Research." *Current Pharmaceutical Design* 6 (2000), 991–1007.

C. K. Van Pelt et al. "A Four-Column Parallel Chromatography System for Isocratic or Gradient LC/MS Analyses." *Analytical Chemistry* 73 (2001), 582–588.

MASS SPECTROMETRY	1961	
Atlas-MAT introduces the first residual gas analyzer based on the quadrupole design.	Potassium-argon dating methods are used to determine that Minnesota rocks are from the Precambrian era.	

HISTORY	1961		
Yuri Gagarin becomes the first person to travel in space.	The U.S.-backed Bay of Pigs invasion fails to overthrow Cuban dictator Fidel Castro.	East and West Berlin are divided by the infamous Berlin Wall.	

K. J. Volk et al. "Profiling Degradants of Paclitaxel Using Liquid Chromatography–Mass Spectrometry and Liquid Chromatography–Tandem Mass Spectrometry Substructural Techniques." *Journal of Chromatography* B 696 (1997), 99–115.

C. M. Whitehouse et al. "Electrospray Interface for Liquid Chromatographs and Mass Spectrometers." *Analytical Chemistry* 57 (1985), 675–679.

H. Zhang et al. "Application of Atmospheric Pressure Ionization Time-of-Flight Mass Spectrometry Coupled with Liquid Chromatography for the Characterization of In Vitro Drug Metabolites." *Analytical Chemistry* 72 (2000), 3342–3348.

The first proceedings of Committee E-14 are published for the ninth annual conference by ASTM.

Chapter 7
7

From Earth to the Planets

Mass spectrometry has been used to explore the atmosphere of Earth and the other planets in our solar system. Most of this research is concerned with determining the constituents of Earth's atmosphere. A requisite characteristic of mass spectrometry in these applications is that in most cases the instrument goes to the sample rather than the reverse. Thus, deployment and operation of the instrument is governed by the restrictions imposed by remote sensing requirements.

The application of mass spectrometry in this area has a long and exciting history. Because mass spectrometers can often provide unambiguous information about the constituents of the atmosphere and the relative abundance of isotopes on Earth and other planets, it is an invaluable tool for both planetary scientists and cosmologists. Although the Annual Meeting on Mass Spectrometry and Allied Topics was initially a venue for presentation of this work, in later years most of it was presented at meetings of the American Geophysical Union or the National Aeronautics and Space Administration (NASA). In any case the results of work in this field are so important and far-reaching that no overview of mass spectrometry would be complete if this subject were ignored. This chapter examines the critical differences imposed by remote sensing, reviews some historic experiments, and discusses some of the most important findings of recent atmospheric and space research with mass spectrometers.

Adapting Mass Spectrometers to a Harsh Environment

One of the two Viking spacecraft that ferried mass spectrometers to Mars to examine the Martian soil.

Early mass spectrometers were large and bulky. When preparing to explore the atmosphere of Earth and the other planets, scientists and engineers were confronted with the nearly

MASS SPECTROMETRY	1962		
	Associated Electrical Industries delivers its MS-9 mass spectrometer to Shell Laboratories in Amsterdam.	A secondary ion mass spectrometry ion microscope is described by researchers in France.	Mass spectrometry is first used at MIT to study the structure of nucleosides.

HISTORY	1962		
	Rachel Carson publishes *Silent Spring*, which ultimately leads to the banning of DDT in the United States.	The world narrowly escapes nuclear war during the Cuban missile crisis.	The Vatican II Council reinterprets the Roman Catholic Church's view of its role in the modern world.

intractable problem of miniaturizing the instrument. Besides reducing its weight and size, they also needed to reduce power consumption and make it rugged enough to withstand acceleration during rocket flight and erratic balloon movements. Remote operation was another requirement. This limitation was especially challenging in the 1950s, when miniaturization of electronic circuits was still in its infancy. Opening and closing of inlets, sampling of gases, control of electronics, data acquisition, and transmission of data to Earth all had to be handled automatically in carefully scripted procedures. Moreover, instrument calibration, that is, setting the intensity and mass scales, had to be done before deployment. The mass spectrometer had to perform reproducibly, since data acquisition at another planet could occur years after instrument calibration prior to launch. Moreover, mass spectrometers sent to other planets had to perform under the harshest of environments—temperatures in excess of 500 degrees Celsius and pressures as high as 100 atmospheres.

Earth's atmosphere is made up of neutral and ionized gases, often interspersed within layers of varying densities. Sampling these regions placed severe constraints on ion source design and operation, since exposure to the ionic constituents of the atmosphere changed rapidly during flight. An early innovation to accommodate this sampling difficulty was the development of the dual ion source, capable of analyzing neutral molecules or ionic species. This source could be switched to accept either neutral or ionic species depending on the region of the atmosphere through which it was passing.

Finally, vacuum system design represented a major challenge for instrument builders. Low-power, lightweight vacuum systems had to perform over wide extremes of pressure outside the instrument. Operational constraints and difficulties notwithstanding, mass spectrometers promised to deliver a wealth of data about our own planet and other planets in our solar system that other analytical techniques simply could not match. Ground-based telescopes, for example, were often limited by interference from Earth's own atmosphere. Even common air pollution could wreak havoc on Earth-based observational experiments designed to measure the components of planetary atmospheres.

A diverse assortment of mass analyzers was available for atmospheric and planetary studies. Magnetic sector, quadrupole mass filters, and time-of-flight (TOF) mass analyzers have all been used for atmospheric and space research. Depending on weight limitations, magnetic sector instruments generally provided higher resolving power, while the quadrupole and TOF mass analyzers were of primary value in studies requiring rapid and continuous scanning, as in cases of rapid passage through the atmosphere. These latter mass analyzers were also more amenable to rapid switching between positive and negative ionization modes.

Two major U.S. government agencies drove the development of mass spectrometry for atmospheric studies after World War II: the U.S. Air Force and NASA. The air force played a significant role in researching Earth's atmosphere. The motivating force behind this research was the need to understand more clearly the composition of the various regions of the atmosphere for improved radio communications and satellite tracking. NASA became

MASS SPECTROMETRY

1963

Early deployment of mass spectrometers occurs on satellites for the study of atmospheric density.

HISTORY

1962

Algeria gains independence from France.

The Beatles score their first hit (in the United Kingdom) with "Love Me Do."

1963

Soviet cosmonaut Valentina Tereshkova becomes the first woman to travel in space.

the central coordinating agency for America's manned spaceflight and planetary research programs, supporting major efforts to determine the origin and evolution of the solar system and to investigate the possibility of life on other planets. NASA-supported studies in mass spectrometry were geared toward acquiring knowledge for its own sake.

Studies of Planet Earth

The use of mass spectrometers to analyze the constituents and structure of Earth's atmosphere has provided a significant increase in our knowledge of this important aspect of our home planet. It has long been known that the lower atmosphere is composed primarily of nitrogen and oxygen and trace amounts of other gases. However, the atmosphere extends from Earth's surface to the border of interplanetary space. Our atmosphere is not a simple diminution of pressure with altitude; it is actually a series of layers or regions, each with its own characteristic constituents. The various regions that have become better defined over the last half century of exploration are the troposphere, stratosphere, mesosphere, thermosphere, and exosphere. Mass spectrometric studies of these regions have led to important scientific discoveries.

Earth, as photographed from the Apollo 17 spacecraft.

In 1961 NASA put into orbit one of the first miniature double-focusing mass spectrographs to study molecules, ions, and free radicals in the exosphere. Designed and built by Consolidated Electrodynamics Corporation, this instrument recorded data continuously during a one-year period at a maximum altitude of 600 kilometers above Earth's surface. Scientists were especially interested in the effects of highly energetic solar and cosmic particles on the gases at the outer edges of our atmosphere. Meanwhile, other missions flown in the early 1960s by the air force collected data on the ionic composition of the atmosphere at altitudes between 64 and 112 kilometers. Whereas most of Earth's atmosphere is composed of neutral molecules, this region, known as the thermosphere, contains ionic species and thus was especially interesting to researchers.

These early mass spectrometric studies discovered previously unidentified ionic species in the thermosphere. In the region above 80 kilometers mass spectral data identified the existence of hydrated water clusters as indicated by the presence of mass peaks at 19 daltons (H_3O^+) and 37 daltons ((H_2O)H_3O^+). While some scientists debated the interpretation of these data, claiming that they resulted from water contamination in the mass spectrometer, the hydrated water cluster explanation was validated several years later when a mechanism for the formation of these and other water-based ions in the lower regions of the thermosphere was confirmed.

It was also during this early mission that NO^+ and O_2^+ and trace concentrations of sodium, magnesium, and calcium ions were recorded, confirming observations made by earlier Soviet studies. Measurements recorded during American missions showed that the metal ions originated from burning meteors as they passed through the atmosphere. The concentration of these metal ions was shown to vary with the frequency of meteoric activity.

Audiocassette tapes are introduced.

Martin Luther King, Jr., delivers his "I have a dream" speech during a massive civil rights march and rally in Washington, D.C.

President John F. Kennedy is assassinated.

1964

The Civil Rights Act bans many forms of institutional racism.

Mass Spectrometry in the Ocean

The need to monitor air quality in submarines has existed since submersibles were first put into service. This need became even more urgent when nuclear-powered submarines joined the U.S. Navy's fleet after World War II. Since these ships typically remain submerged for periods up to three months, the buildup of pollutants from shipboard materials and respiration products of the crew needed to be carefully monitored.

The Naval Research Laboratory (NRL) was charged with the responsibility of devising a means of monitoring shipboard atmospheres in submarines. To maintain optimal air quality, it was deemed necessary to measure the levels of eight gases in submarine atmospheres: nitrogen, oxygen, hydrogen, water vapor, carbon monoxide, carbon dioxide, and two freons, R-12 and R-114. As a first approach a series of different sensors, each sensitive to one or more of the gases of interest, was chosen. However, the requirement of a monitoring system that performed reliably around the clock with minimal operator intervention posed an insurmountable problem for these early multisensor systems.

While mass spectrometers had been used on board ships in earlier experiments under the supervision of technically competent staff, they were considered too complex for unattended operation and had the additional difficulty that they could only accurately measure four of the eight gases required: nitrogen, oxygen, carbon dioxide, and water vapor. Another problem was that carbon monoxide and nitrogen have the same mass, so without a high-resolution instrument the very important gas, carbon monoxide, could not be monitored. However, development of reliable, miniaturized mass spectrometers for space exploration and the need to have an atmospheric monitoring device on board submarines fostered the development of better instruments. A series of instruments was designed, built, and tested for this purpose, ranging from single magnetic-sector mass spectrographs to double-focusing sector instruments to quadrupole mass filters. In the first trial the four gases listed above were monitored, and a separate device was used to monitor the concentration of carbon monoxide. The trial was a success in that the instrument performed with minimal attention for over seventy days, much longer than any previous monitoring devices. On the basis of this success the NRL tackled the problems associated with monitoring the other gases and the freons. Eventually, a combination mass spectrometer–infrared atmosphere monitor was able to provide accurate real-time analyses of the atmospheric constituents reliably throughout a typical ninety-day submarine mission.

In the mid-1970s NASA began a research program using the Atmosphere Explorer (AE) satellites to study in detail the physics and chemistry of the high-altitude regions of Earth's atmosphere. Three satellites were launched, one of which contained a double-focusing mass spectrometer assembled by Alfred Nier and his research group at the University of Minnesota. A significant discovery during this mission was the presence of atomic nitrogen (^{14}N) at high altitudes. Although its presence had been predicted earlier and verified qualitatively by optical spectroscopic techniques, it was not until Nier's instrument passed through the upper atmosphere that precise quantitative measurements were taken. Evidence for the presence of atomic nitrogen was inferred from the observation of a peak at mass 30—nitric oxide (NO)—at altitudes of 400 kilometers and higher. The intensity of the NO peak was much too great for the known ambient concentration of NO at this altitude, thus prompting researchers to conclude that nitrogen atoms were colliding with atomic oxygen

MASS SPECTROMETRY	1964		
	The concept of a computer search of mass spectra is first proposed.	The flowing afterglow technique for gas-phase ion-molecule reaction studies is developed at the U.S. National Bureau of Standards.	A jet separator for GC-MS is demonstrated by Swedish chemist Ragnar Ryhage.

HISTORY	1964		
	Nikita Khrushchev is ousted from power in the Soviet Union in a hard-line coup led by Leonid Brezhnev.	Murray Glen-Mann proposes that matter is made of quarks.	The Gulf of Tonkin Resolution authorizes full U.S. military intervention in Vietnam.

on the surfaces of the instrument to form NO, which was subsequently detected by the mass spectrometer. Direct confirmation of atomic nitrogen at lower altitudes was obtained when the intensity of the ion current at 14 daltons was greater than that observed in the reference spectrum of molecular nitrogen. As Nier himself recalled a decade later: "This study, together with a series of subsequent investigations, established many interesting facts about the behavior of atomic nitrogen in the earth's upper atmosphere."

Beginning in the late 1950s, an airborne mass spectrometry program was organized at the Air Force Research Laboratory (AFRL) near Boston, Massachusetts. The purpose of this new program was to study the behavior of the ionosphere, about which very little was known. The ionosphere controls radio-wave propagation. As innovations in radio technology pushed transmission signals further into the microwave region, studies of this nature became increasingly important.

Scientists and engineers at AFRL directed their efforts toward the development of a compact and rugged unit based on the new quadrupole mass filter design recently introduced by German physicist Wolfgang Paul. In addition to examining the types of ions in the ionosphere and how they affected radio transmission and propagation, these early mass spectrometers gathered data on ion production and neutralization mechanisms. On the ground, researchers in the laboratory used these data to study ion kinetics. Throughout the 1960s improved quadrupole mass filters were included on rocket and satellite payloads. Studies were conducted during meteor showers, auroras, and solar eclipses and at latitudinal variations.

Using rockets as platforms for atmospheric studies has two major disadvantages: flights are of relatively short duration and are one-shot experiments. Satellite-based mass spectrometers can take measurements over much longer periods. Capitalizing on this advantage, AFRL has conducted more than 350 satellite experiments since 1958. A large portion of this work dealt with improving our knowledge of the atmospheric density. One of the primary responsibilities of the air force is to track the thousands of satellites and pieces of space debris that orbit Earth. Orbital predictions and re-entry probabilities are calculated at the North American Aerospace Defense Command (NORAD) using algorithms that employ computer modeling of atmospheric density to determine the frictional drag on satellites. Mass spectrometers in low-Earth orbits have been used to measure the absolute density of the atmosphere. When correlated with the position, time, and solar geomagnetic activity, these measurements validated and improved data on which the models were based.

Most of the mass spectrometric investigations carried out during the 1960s involved positive ion species. Measurements of negative ion composition in lower regions of the atmosphere proved much more challenging. Consider, for example, a rocket-borne mass spectrometer traveling at high speed. The entire rocket is at a slight negative potential owing to its interactions with the ambient plasma. This effect produces an electric field that repels negative ions. Unfortunately, the simple expedient of using a positive voltage to draw negative ions into the mass spectrometer also attracts electrons. Since electrons make up the

1965

Researchers at MIT describe the molecular effusion GC-MS interface.

1965

An electrical blackout grips the northeastern United States.

Malcolm X is assassinated during a speech at the Audubon Ballroom in New York City.

Riots sweep through the poor and largely African-American Watts neighborhood of Los Angeles.

largest component of the negatively charged species in this region, the strong electron current interferes with the measurement of the negative ions of interest. To overcome this problem, modifications had to be made to the sampling geometry and to the detection electronics of airborne mass spectrometers. Because of the low concentration of the negatively charged species, ordinary detectors were not sensitive enough to register these species. Consequently, more sensitive pulse-counting technology was used for these measurements. As noted earlier, TOF and quadrupole designs permit rapid switching between positive and negative ion modes of operation, so they are most frequently used in experiments in which ions of both polarities are to be monitored.

Equally significant was the pioneering research AFRL conducted on very heavy ions previously thought not to exist in the atmosphere. In the late 1960s researchers had detected ions with a mass range in excess of 150 daltons at altitudes between 80 and 90 kilometers. More than a decade later, in 1982, additional experiments confirmed the existence of very large ions with weights in excess of 450 daltons. Although the origins are still unclear, one theory suggests that these large ions are extremely small meteoric dust particles that have become multiply charged by electron capture processes.

From Meteorites to the Moon

In many cases fragments of meteorites are found on Earth's surface, where they can be collected and easily transported to the laboratory for mass spectral analysis. Similarly, lunar samples retrieved by the manned Apollo missions to the Moon during the late 1960s and early 1970s provided researchers with the unique opportunity to apply mass spectrometry to study the origins and characteristics of Earth's nearest neighbor.

View of the Moon photographed from the Apollo 11 *spacecraft during its journey homeward.*

Meteorites have provided a rich source of information about our solar system and the space around us. Most meteorites are of the stony-iron type. Here again, isotope ratio mass spectrometry techniques can be used to determine meteorite age and the salient features of a meteorite's history. Much rarer are the class of meteorites called carbonaceous chondrites. These meteorites contain small amounts of carbon in the form of organic compounds. Only a handful have fallen in the last several decades, and the organic content of these specimens has been the subject of intensive scrutiny by geologists using a variety of analytical tools, including mass spectrometry.

Early mass spectral analysis showed that common hydrocarbon compounds, such as simple waxes and benzenes, were indigenous to these meteorites. Perhaps the most startling discovery was the later observation of amino acids in the Murchison carbonaceous chondrite. This result was corroborated by other researchers in 1976 who studied the Murchison and Murray meteorites. Not only were some of the essential amino acids for life found, but their optical isomers were also present in nearly equal proportions. Several non-essential amino acids were detected as well. The fact that the amino acids are present in both optical isomeric forms is compelling evidence that these compounds are present in carbonaceous

MASS SPECTROMETRY	1966	
	Tandem mass spectrometry for ion-molecule reaction studies is developed at Aerospace Research Laboratories, Wright-Patterson Air Force Base.	Researchers at Stanford University describe an ion-cyclotron double resonance instrument.

HISTORY	1965	1966	
	Cosmic microwave radiation is discovered by Arno H. Penzias and Robert W. Wilson and is thought to be a residual effect of the big bang.	China's Cultural Revolution begins under the direction of Mao Zedong, throwing nearly all aspects of Chinese life into chaos.	Miranda rules are established by the U.S. Supreme Court.

Astronauts on NASA's Apollo missions to the Moon brought lunar samples back to Earth for analysis by mass spectrometry.

chondrites and are not just the result of contamination, a constant concern among scientists who do this type of research. These findings may have important implications for some theories regarding the origin of life on Earth.

Between 1969 and 1972 NASA dispatched six manned missions to the Moon. Soil and rock samples were collected and returned to Earth for analysis. Researchers focused their efforts on determining the presence and types of organic compounds and measuring the relative abundances of isotopes to further their understanding of the age of the lunar samples. Gas chromatography mass spectrometry (GC-MS) was the most sensitive method for organic content studies, capable of isolating organic compounds at the level of one part per billion. However, when contamination factors were eliminated from the analyses, no evidence was obtained to suggest the existence of life on the Moon or even the presence of organic materials. Nevertheless, isotope ratio mass spectrometric studies of the inorganic constituents of lunar samples did yield more useful results. Isotopic analysis of rubidium-strontium levels in rocks estimated their age at more than three-and-a-half billion years. Similar examinations placed the age of the lunar soil at four-and-a-half billion years. Prevailing theory claims that the soil was formed from the primordial lunar material and that some of the rocks formed later during heavy volcanic activity on the lunar surface.

Planets and Comets

Studies of Earth's atmosphere and materials found on the lunar surface were complemented by American and Soviet efforts to explore other planets in the solar system. Venus was the first planet selected for study. In the early 1960s little was known about the Venusian atmosphere beyond general knowledge that it contained carbon dioxide and trace amounts of other gases. According to Alfred Nier, "Improved infrared astronomy and radiotelescope data greatly enhanced our knowledge, but it was not until 1962 when *Mariner 2* flew by the planet and shortly thereafter, when the Soviet Venera series of spacecraft began to land on the planet, that our knowledge made substantial strides." The launch of *Venera 4* in 1967 marked the first of several missions to study the Venusian atmosphere, while later missions surveyed the planet's surface. Data transmitted back to Earth indicated that Venus had a very harsh environment indeed: a dense atmosphere consisting mostly of carbon dioxide and

1967

Chemical ionization mass spectrometry is developed at the Esso Research and Engineering Company.

Researchers at MIT use a computer to sequence oligopeptides from their spectra.

Alternation of accelerating voltage for selected ion monitoring in magnetic sector mass spectrometers permits resolution of mixtures in GC-MS analysis.

A reverse geometry sector instrument is marketed by Mess- und Analysen- Technik (MAT).

1967

Suharto comes to power in Indonesia.

Jocelyn Bell discovers pulsars.

The Arab-Israeli Six-Day War ends. Israel occupies the West Bank, Golan Heights, Sinai Peninsula, and East Jerusalem.

Ultraviolet image of Venus's clouds as seen by the Pioneer Venus orbiter, 26 February 1979.

nitrogen generated extreme conditions on the surface. Pressures nearly one hundred times greater than those experienced on Earth were common, while daytime temperatures soared to 500 °C. Clearly, these severe conditions would test future missions equipped with sensitive analytical instruments. During the next fifteen years both the United States and the U.S.S.R. flew increasingly sophisticated missions to investigate the Venusian atmosphere in greater detail. In 1975 *Venera 9* sampled the Venusian atmosphere with a mass spectrometer, providing the first detailed information about the constituents of its upper regions. Neutral molecules, such as carbon monoxide, ammonia, carbonyl sulfide, carbon dioxide, and carbon disulfide, were detected down to an altitude of 63 kilometers above the planet's surface. Positive ions of carbon dioxide and atomic nitrogen were also observed at the same altitude.

NASA launched the first *Pioneer* spacecraft to Venus in 1978. *Pioneer* was a veritable laboratory of mass spectrometers (see the Table) The Venus bus carried the mission components during transit from Earth to Venus, while the Venus orbiter served as the platform from which to launch the probes. All of these separate units contained mass spectrometers.

Table. Summary of Mass Spectrometers on Pioneer 13 Mission to Venus

Pioneer component	Instrument	Mass analyzer	Measurement	Mass (kg)	Power (W)	Mass range (Da)
Venus bus	Ion mass spectrometer	Bennett radio frequency	Solar wind interaction, upper atmosphere photochemistry	1	1	1–60
	Neutral mass spectrometer	Magnetic sector, double focusing	Origin and evolution of Venusian atmosphere	5	6	1–46
Venus orbiter	Ion mass spectrometer	Bennett radio frequency	Concentration of thermal positive ions in Venusian ionosphere	3	1.5	1–60
	Neutral mass spectrometer	Quadrupole mass filter	Density variations of the major constituents of the upper atmosphere	3.8	12	1–45
Venus large probe	Neutral mass spectrometer	Magnetic sector, double focusing	Composition of the lower atmosphere	9	12	1–212

Source: NASA Web page. http://nssdc.gstc.nasa.gov/planetary/pndata.html.

MASS SPECTROMETRY

1968

The electrospray ionization technique is introduced at Northwestern University for studying macromolecules.

HISTORY

1967

Christiaan Barnard performs the first successful human heart transplant on Louis Washansky.

China explodes its first hydrogen bomb.

1968

Nationwide student and labor protests paralyze France.

Artist's conception of the Pioneer Venus orbiter as it circles Venus.

The data collected by the instruments described in the Table enabled scientists to compile a detailed picture of the lower and upper portions of the Venusian atmosphere. Carbon dioxide (96.5 percent) and nitrogen (3.5 percent) made up most of the near-surface atmosphere. The sensitive mass spectrometers also measured minor constituents with concentrations on the order of one part per million (ppm). Sulfur dioxide (150 ppm), argon (70 ppm), water (20 ppm), carbon monoxide (17 ppm), helium (12 ppm), and neon (7 ppm) were among the components identified and quantified.

Meanwhile, measurements in the upper atmosphere unequivocally indicated the presence of atomic oxygen ions and atomic carbon in the topside of the ionosphere, while nitrogen, hydrogen, and helium ions appeared as lesser constituents. At lower levels molecular oxygen, NO^+, carbon monoxide, and carbon dioxide ions were dominant. During the Pioneer missions, it was determined that, just like Earth, the atmosphere of Venus is dynamic and responsive to the solar wind. Subsequent studies of the Venusian atmosphere during a two-year period enabled researchers to obtain a complete picture of the behavior of the atmosphere under a wide variety of conditions. This work showed significant diurnal variations of the positive ion species, modified by both sunspot activity and changes in the solar wind.

Successive missions by the Americans and Soviets moved beyond interest in the atmospheric constituents to the more subtle experiment of measuring the relative abundances of

The order of gas phase basicity of neutral methylamines is determined by ion cyclotron resonance-mass spectrometry.

Researchers in Australia use a computer library search to identify spectra of unknown compounds.

Martin Luther King, Jr., is assassinated.

Robert F. Kennedy is assassinated.

The Tet Offensive is launched by North Vietnamese forces.

Top: During the Viking 1 mission to Mars, soil samples taken by the collector shown here were run through a gas chromatograph mass spectrometer.

Bottom: Members of the Viking mission team examine data returned from the entry science neutral-atmospheric composition mass spectrometer.

the isotopes of the atmospheric gases. Data obtained during the measurement of inert gases in the Venusian atmosphere have been interpreted differently by different scientists. These gases—helium, neon, argon, krypton, xenon, and radon—are chemically inert and do not react with other atmospheric components. Thus, data about the abundance of the various isotopes are assumed to reflect their primordial concentration and can be used to speculate about the origin of the Venusian atmosphere.

Soviet scientists interpreted data from the Venera missions to indicate that Venus and Earth experienced a similar history in the accretion of their atmospheres. Scientists in America, however, studied the data on krypton and xenon gathered during the Pioneer missions and concluded that the Venusian atmosphere accumulated these noble gases as a result of the continual bombardment of the planet's atmosphere by the solar wind. These missions also produced evidence of a high concentration of deuterium ions in the atmosphere. When this information was fed into atmospheric models, the results suggested that Venus may have supported a large water ocean early in its history.

While Venus has a harsh environment that is apparently unable to support life as we know it, Mars, according to some scientists, appeared to be a better candidate for the existence of living organisms. Even before the first probes were sent to Mars in the 1960s, scientists were familiar with some of the red planet's features, such as atmospheric density and temperature extremes on the surface. The search for life, however, was one of the most compelling reasons for NASA's vigorous efforts to land a spacecraft on the Martian surface in the 1970s. GC-MS played an important role in this endeavor.

Project Viking got under way in 1975, when two spacecraft departed NASA's Cape Canaveral launch facility on the Florida coast. Each mission contained specialized mass spectrometers; one to assay the thin Martian atmosphere during descent and another to analyze surface materials in search of organic substances that would be indicative of the existence of life. At the University of Minnesota, Alfred Nier built the double-focusing mass spectrometer for atmospheric analysis during the Mars mission. The data obtained from this instrument indicated that the major constituents of the Martian atmosphere are carbon dioxide (95.32 percent), nitrogen (2.7 percent), argon (1.6 percent), oxygen (0.13 percent), and carbon monoxide (0.08 percent). The high concentration of carbon dioxide, which matched closely the levels found on Venus, and the very low atmospheric pressure diminished scientists' hopes that life existed on Mars.

The likelihood of life on Mars was further reduced by results from mass spectral analysis of Martian soil samples in a series of GC-MS experiments orchestrated by scientists at MIT. Despite the instrument's very high sensitivity for organic compounds—several parts per billion by weight—none were identified. The soil analysis mass spectrometer was subsequently used to sample and concentrate the atmosphere for an accurate determination of the $^{15}N/^{14}N$ ratio. The ratio was determined to be 60 percent higher than that on Earth, indicating that the Martian atmosphere must have contained significantly more nitrogen early in its history.

MASS SPECTROMETRY		1969	
		The first gas chromatograph mass spectrometer with an integrated computer data system is introduced.	The American Society for Mass Spectrometry (ASMS) is incorporated on August 8.

HISTORY	1968	1969	
	Czechoslovak leader Alexander Dubček's democratic reforms, known as the Prague Spring, provoke a Soviet invasion that places a hard-line regime in power.	The Cuyahoga River catches fire in Cleveland owing to the presence of high levels of flammable pollutants.	The Woodstock festival takes place.

Jupiter and its four planet-sized moons photographed by Voyager 1 *and assembled into a collage. The moons are not to scale, but they are in their relative positions.*

Right: The Martian landscape as seen from the Viking 2 *lander in November 1976.*

Some scientists suggest that a catastrophic event in the distant past removed more than 90 percent of the Martian atmosphere.

Venus and Mars are smaller and more similar to Earth than is Jupiter, a large gaseous planet that may not even have a hard surface. Although it is the largest planet in the solar system, Jupiter is not massive enough to support the nuclear reactions that fuel stars like our Sun. Several missions were flown to Jupiter before an atmospheric entry probe was sent on the Project Galileo spacecraft in 1989. After a six-year journey, the probe, equipped with a quadrupole mass filter, entered and sampled the Jovian atmosphere during descent. Data, telemetered back to the Galileo satellite before the atmospheric entry probe was destroyed by the harsh environment at lower altitudes, provided a detailed analysis of the atmosphere. The most abundant gases were found to be helium and hydrogen, with a ratio slightly less than the solar value. On the basis of this result scientists believe that at lower altitudes, a helium "rain" is very slowly reducing the content of that element in the upper regions of the atmosphere. The probe also detected methane, water, ammonia, hydrogen sulfide, deuterium, and the noble gases—neon, argon, krypton, and xenon.

Mass spectrometers have also been used to investigate even more ephemeral bodies than gaseous planets, namely, comets. The European-sponsored Giotto mission was designed to study Comet P/Halley by means of a close flyby. Two of the major objectives of the mission were to determine the elemental and isotopic composition of volatile components in the coma, particularly parent molecules, and to determine the elemental and isotopic composition of dust particles. Both of these tasks were accomplished by mass spectrometers during its encounter with Halley in March 1986, at a distance of 0.89 astronomical units (AUs)

1970

The annual Conference on Mass Spectrometry and Allied Topics becomes a conference of the ASMS, initially in cooperation with the ASTM Committee E-14.

Development of algorithms is begun for computer-based comparison of mass spectral data to libraries of known spectra for automated identification of unknown compounds.

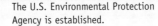

1970

Apollo 11 lands on the Moon.

The U.S. Environmental Protection Agency is established.

Four students are killed when National Guard troops fire on Vietnam War protesters at Kent State University in Ohio.

Galileo Timeline

The Galileo probe became the first atmospheric probe to examine a gas giant on 7 December 1995. The key events that occurred are listed below. Times are given in Eastern Standard Time.

Time	Event
11:04 A.M.	Coast timer initiates probe operation.
12:46 P.M.	Orbiter flies by Io (~1000 km) (no imaging or spectral data collected).
2:04 P.M.	Energetic Particles Investigation begins, measuring trapped radiation in a region previously unexplored.
4:54 P.M.	Orbiter reaches closest point to Jupiter.
5:04 P.M.	Probe enters Jupiter's atmosphere and data relay begins.
5:05:52 P.M.	Pilot parachute deploys.
5:05:54 P.M.	Main parachute deploys.
5:06:02 P.M.	Deceleration module is jettisoned.
5:06:06 P.M.	Direct scientific measurements begin.
5:06:15 P.M.	Radio transmission to orbiter begins.
~5:08 P.M.	Probe reaches visible cloud tops of Jupiter.
5:12 P.M.	Atmospheric pressure is recorded (it is the same as Earth's sea-level pressure).
5:17 P.M.	Second major cloud deck is encountered (uncertain).
5:28 P.M.	Water clouds are entered (uncertain).
5:34 P.M.	Atmospheric temperature is recorded (it is equal to room temperature on Earth).
5:46 P.M.	Probe enters twilight.
6:04 P.M.	Baseline mission ends. Probe may cease to operate because it loses battery power or is crushed, or atmosphere may weaken signal.
6:19 P.M.	Orbiter ceases to receive probe data (if, in fact, it is still transmitting).
7:27 P.M.	Galileo main engine is ignited (remains active for 49 minutes) to insert into Jovian orbit.

Note that data acquisition in the Jovian atmosphere lasted barely one hour. It is estimated that by 3:00 A.M. (EST) on 8 December 1995, the probe had been completely destroyed by the heat in Jupiter's atmosphere.

from the Sun and 0.98 AUs from Earth. The closest approach to Halley was 596 kilometers. Despite a brief, four-hour encounter with the comet and the loss of some instrumentation owing to a "large particle" impact, the mission was a success, and for the first time direct measurement of compounds in the coma was accomplished. In addition to water, evidence for the presence of acetylene, formaldehyde, and hydrogen cyanide was found.

MASS SPECTROMETRY	1970		1971
	NASA researchers discover amino acids in carbonaceous chondrites.	Eiji Osawa at Hokkaido University proposes a carbon compound with a three-dimensional "soccer ball" structure.	Glow discharge mass spectrometry is developed at IBM research laboratories in San Jose, California.

HISTORY	1970		1971
	A cyclone leaves between 150,000 and 200,000 dead in East Pakistan (now Bangladesh).	U.S. forces withdraw from Cambodia.	Bangladesh gains independence from Pakistan.

Artist's conception of the Galileo *spacecraft releasing its atmospheric probe near Jupiter.*

Right: Halley's comet.

As we have seen in this chapter, the application of mass spectrometry to planetary studies is a fascinating and ongoing enterprise. The most recent surprise in this area is the discovery of buckminsterfullerenes—extremely durable carbon structures—in the Pueblito de Allende carbonaceous chondrite. In addition to C_{60}, researchers from the University of Hawaii and NASA have recently observed C_{76} to C_{96} fullerenes as well as an envelope of mass peaks indicative of fullerenes containing from 100 to 250 carbon atoms. Some of these fullerenes have not been observed in any terrestrial studies to date. NASA is currently planning missions to send advanced analytical instrumentation to Mars and Saturn and through the coma of other incoming comets. Ongoing commitments to the study of our home planet and planetary system will continue to include mass spectrometry as an important analytical tool to increase our knowledge of Earth, our solar system, and the universe.

Suggested Reading

P. I. Abell et al. "Organic Analysis of the Returned Apollo 11 Lunar Sample."In *Proceedings of the Apollo 11 [Eleven] Lunar Science Conference*, 1757–1773. New York: Pergamon Press, 1970.

E. C. Alexander, Jr.; P. K. Davis; J. H. Reynolds. "Rare-Gas Analyses on Neutron Irradiated Apollo 12 Samples." In *Proceedings of the 3rd Lunar Science Conference*,Vol. 2, 1787–1795. Cambridge, Mass.: MIT.

L. Becker; T. E. Bunch; L. J. Allamandola. "Higher Fullerenes in the Allende Meteorite." *Nature* 400 (1999), 227–228.

K. Biemann. "Organic Analysis." *Applied Optics* 9 (1970), 1282–1288.

"Static" secondary ion mass spectrometry is developed at the University of Cologne.

GC-MS is used for clinical diagnosis of metabolic disorders.

Reflectron time-of-flight mass spectrometry is developed in Leningrad.

1972

Attica Prison is taken over by inmates.

The Soviets launch *Salyut 1*, the first space station.

Democratic National Committee headquarters in the Watergate Hotel is burglarized.

K. Biemann et al. "Search for Organic and Volatile Inorganic Compounds in Two Surface Samples from the Chryse Planitia Region of Mars." *Science* 194 (1976), 72–76.

A. L. Burlingame et al. "Study of Carbon Compounds in Apollo 11 Lunar Samples." In *Proceedings of the Apollo 11 [Eleven] Lunar Science Conference*, 1779–1791. New York: Pergamon Press, 1970.

T. M. Donahue; J. H. Hoffman; R. R. Hodges, Jr. "Krypton and Xenon in the Atmosphere of Venus." *Geophysical Research Letters* 8 (1981), 513–516.

F. C. Fehsenfeld; E. E. Ferguson. "Origin of Water Clusters in the D Region." *Journal of Geophysical Research* 74 (1969), 2217–2222.

J. M. Gibert; J. Oro. "Gas Chromatographic-Mass Spectrometric Determination of Potential Contaminant Hydrocarbons of Moon Samples." *Journal of Chromatographic Science* 8 (1970), 295–296.

R. E. Hartle; H. A. Taylor, Jr. "Identification of Deuterium Ions in the Ionosphere of Venus." *Geophysical Research Letters* 10 (1983), 965–968.

P. M. Hurley; W. H. Pinson, Jr. "Whole-Rock Rubidium-Strontium Isotopic Age Relationships in Apollo 11 Lunar Samples." In *Proceedings of the Apollo 11 [Eleven] Lunar Science Conference*, 1311–1315. New York: Pergamon Press, 1970.

V. G. Istomin. "Ions of Extraterrestrial Origin in the Earth's Ionosphere." *Space Research* 3 (1963), 209–220.

M. N. Izakov. "Noble Gases in the Atmosphere of Venus, Earth, and Mars: The Problem of the Origin of Planetary Atmospheres." *U.S.S.R. Institute of Space Research Report* (1979), 1–27.

G. Jungclaus et al. "Aliphatic Amines in the Murchison Meteorite." *Nature* 261 (1976), 126–128.

F. R. Krueger. "Carbonaceous Matter in Cometary Dust and Coma." *Advances in Space Research* 15 (1994), 407–411.

K. A. Kvenvolden et al. "Evidence for Extraterrestrial Amino Acids and Hydrocarbons in the Murchison Meteorite." *Nature* 228 (1970), 923–926.

"Mass Spectrometer Going into Orbit." *Electronics* (26 Feb. 1960), 81.

M. B. McElroy et al. "Composition and Structure of the Martian Atmosphere: Analysis of Results from Viking." *Science* 194 (1976), 1295–1298.

R. C. Murphy et al. "Search for Organic Material in Lunar Fines by Mass Spectrometry." In *Proceedings of the Apollo 11 [Eleven] Lunar Science Conference*, 1891–1900. New York: Pergamon Press, 1970.

R. S. Narcisi; A. D. Bailey. "Mass Spectrometric Measurement of Positive Ions at Altitudes from 64 to 112 Kilometers." *Journal of Geophysical Research* 70 (1965), 3687–3700.

R. S. Narcisi et al. "Processes Associated with Metal-Ion Layers in the E Region of the Ionosphere." *Space Research* 8 (1968), 360–369.

H. B. Niemann et al. "The Galileo Probe Mass Spectrometer: Composition of Jupiter's Atmosphere." *Science* 272 (1996), 846–849.

A. O. Nier. "Mass Spectrometry in Planetary Research." *International Journal of Mass Spectrometry and Ion Processes* 66 (1985), 55–73.

A. O. Nier; J. L. Hayden. "Miniature Mattauch-Herzog Mass Spectrometer for the Investigation of Planetary Atmospheres." *International Journal of Mass Spectrometry and Ion Physics* 6 (1971), 339–346.

Y. A. Surkov et al. "Mass Spectral Study of the Chemical Composition of the Atmosphere of Venus by 'Venera-9' and 'Venera-10' Automated Space Probes." *Geokhimiya* (April 1978), 506–513.

MASS SPECTROMETRY | **1972**

Mass-analyzed ion kinetic energy spectrometry (MIKES) is developed.

HISTORY | **1972**

DDT is banned in the United States.

Eleven Israeli athletes are killed by terrorists at the Olympic Games in Munich.

The United States and the U.S.S.R. sign the Strategic Arms Limitation Treaty (SALT I).

A. A. Viggiano; D. E. Hunton. "Airborne Mass Spectrometers: Four Decades of Atmospheric and Space Research at the Air Force Research Laboratory." *Journal of Mass Spectrometry* 34 (1999), 1108–1111.

D. R. Williams. *Pioneer Venus Project Information.* National Space Science Center. 18 Dec. 2001. <http://nssdc.gsfc.nasa.gov/planetary/factsheet.html>.

Direct Quotations

Page 93, Nier, 1985, p. 63.

Page 95, Nier, 1985, p. 68.

1973

The self-training interpretive and retrieval system (STIRS) for interpretation of mass spectra is introduced.

1973

Egypt and Syria attack Israel, beginning the Yom Kippur War.

An OPEC oil embargo leads to fuel shortages and high gasoline prices in the United States.

U.S. forces withdraw from Vietnam.

Chapter **8** *Environmental Distress*

Right: Figure 5 from Wolfgang Paul's patent of the linear quadrupole mass filter. The quadrupole rods are shown connected to the appropriate DC and RF power supplies.

Left: Exhaust fumes from automobiles have reduced air quality in most urban areas throughout the United States. Shown here is a Southern California freeway in 1972.

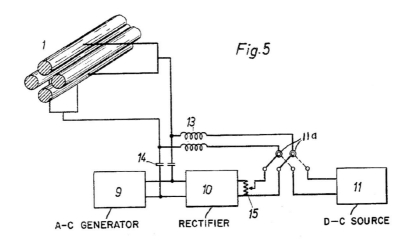

Researchers investigating the broad relationship between human activity and climatic and environmental changes typically work with a large set of complex and often controversial variables. Some scientists, for example, have shown that holes in Earth's ozone layer are indicative of serious damage to the atmosphere, while other experts argue that more evidence needs to be gathered over a longer period before an accurate conclusion can be drawn. Similarly, in the case of dioxins, researchers have yet to agree on a satisfactory explanation of their impact on human health and the natural environment. For more than thirty years the mass spectrometer has helped eliminate some of the scientific uncertainty in an important field of investigation that is still very contentious. Mass spectrometry has provided accurate and timely data for the evaluation of a multitude of environmental contaminants on the local, regional, national, and global levels.

The mass spectrometer is a critical enabling technology for environmental research. It has been used to identify and measure the concentration of many chemicals in soil, water, and air. Environmental scientists have used these data to examine the effect of hazardous material on general health trends and patterns of ecological change. Equally significant is how such studies have shaped public attitudes toward risk assessment and regulation. Mass spectrometers can measure substances in water at levels that were previously undetectable by older analytical methods. Even though concentrations of such substances may be well below

MASS SPECTROMETRY   **1974**

| Fused silica capillary gas chromatography columns are introduced for GC-MS. | Fourier transform-ion cyclotron resonance mass spectrometry is introduced. | An atmospheric pressure chemical ionization interface for LC-MS is developed. |

HISTORY   **1974**

| A Communist coup overthrows Emperor Haile Selassie I of Ethiopia. | The U.S. Freedom of Information Act is passed. | Richard Nixon resigns the presidency. |

Rachel Carson.

harmful levels, the mere fact that they are detected at all raises public awareness and anxiety. The ability of the mass spectrometer to measure increasingly minute quantities of potentially harmful substances on land and in our streams and air may lower the acceptance threshold of what experts and the public at large consider to be "safe" levels of tolerance.

In 1962 ecologist Rachel Carson published *Silent Spring,* which examined the environmental impact of widespread pesticide use. This landmark study helped spark the environmental movement in modern America. Eight years later, under increased public pressure to acknowledge the deterioration of America's environment, President Richard Nixon established by executive order the U.S. Environmental Protection Agency (EPA). Aggressive technological innovation and commercialization of gas chromatography mass spectrometry (GC-MS) in the 1970s made possible the large number of analyses required by the regulations promulgated by the EPA. These new guidelines for the identification and treatment of environmental pollutants, first in water and later on land and in the air, helped create a tremendous and unprecedented demand for new analytical techniques. GC-MS met this analytical demand because of its ability to analyze large quantities of samples rapidly and at low cost. Modern GC-MS instruments were also made easy to operate, especially after the first bench-top units appeared in the mid-1970s. For environmental GC-MS analysis, highly trained personnel were no longer needed to operate instruments. Expertise in instrument control as well as in data acquisition and handling were increasingly built into the instruments themselves.

The Need for Analytical Instrumentation

The history of environmental mass spectrometry is not simply a story of scientific discovery. It is a story about the development and commercialization of technology. Early developments in mass spectrometry were driven by intellectual pursuits to understand the fundamentals of physics and the atomic and molecular composition of matter. The commercial development of mass spectrometers was initially guided by the analytical needs of the petroleum industry and the massive wartime Manhattan Project. After World War II this development was driven further by the increasing number of diverse researchers working in the chemical and later in the pharmaceutical industries who desired more powerful means to measure the abundance and determine the composition of mixtures and elucidate the structure of compounds. Commercial mass spectrometry got a tremendous boost in the 1970s when GC-MS became the instrument of choice for environmental analysis.

Congress mandated the cleanup of America's natural landscape. The legislation that Congress passed enabled the EPA to implement regulatory policies and then to develop guidelines for the accurate identification and precise measurement of environmental pollutants. Complete analytical methods, ranging from sample acquisition and preparation to measurement and data interpretation, were developed, prescribed, and applied to a wide variety of environmentally significant substances—air, wastewater, groundwater, drinking water,

MASS SPECTROMETRY | **1975**

Mass spectrometers are placed on board NASA's *Viking* spacecraft for its mission to Mars.

HISTORY | **1975**

Francisco Franco dies after ruling Spain for thirty-six years.

Saigon falls to North Vietnamese forces.

The first personal computers are introduced.

Environmental Distress

Pulp mills in West Virginia discharge pollutants into the Columbia River, 1972.

soils, sediments, and animal tissue. Analytical instruments had to analyze many samples in a very short period and at low cost. Furthermore, quality control and assurance were emphasized in an attempt to forestall inaccurate results.

The new environmental analyst worked in a profoundly different mode and faced entirely different challenges than his or her academic or industrial counterparts. The new environmental regulations called for analysts to run thousands of samples in order to collect information about numerous "target" substances of concern. In such a large-scale national program, cost control, speed, consistency, and accuracy were of paramount concern. Unlike academic and industrial researchers, most environmental analysts did not have the luxury of deep, specialized knowledge of the chemical problems at issue or of the analytical instruments required to measure every sample. In the case of environmental testing and regulation the scale of the necessary analytical efforts required that experts develop, and the government prescribe, methods that would meet the constraints of speed and cost. The effort would require large laboratories staffed by technicians working with new instrumentation and possessing minimal mass spectrometric expertise.

The creation of the EPA was followed by a series of major legislative reforms directed toward addressing environmental issues in the United States. Between 1970 and 1980 Congress passed acts that mandated improvements of America's air, water, and land. In 1972 Congress passed the Federal Water Pollution Control Act Amendments, which resulted in the EPA's creation of regulations with strict limits on the levels of industrial waste that could be discharged into U.S. waters. These acts mandated rigorous testing for numerous chemicals before their amounts and roles as pollutants were completely known, even before methods and instruments were available for their measurement. In 1976 environmental activist groups sued the EPA for failing to implement some provisions of this act. A consent decree was drawn up to settle the lawsuit. Included in the decree was a priority list of 129 potential pollutants being discharged into streams and lakes through industrial and municipal wastewaters. The EPA developed standard sampling and analysis techniques in conjunction with public input, and these priority compounds and methods were incorporated into subsequent environmental programs. Institutionalization of these and other legislative acts put tremendous

1976

The selected ion flow tube is developed by scientists in Great Britain to enable precise control of reactant ions.

Plasma desorption mass spectrometry and its application to the study of biomolecules is developed at Texas A&M University.

The first bench-top gas chromatograph mass spectrometer is introduced.

Japanese researchers take isotope-ratio measurements using the first combination of gas chromatography combustion mass spectrometry.

1976

The first home VCRs are introduced.

The superstring theory is introduced, originally to explain the origins of the strong nuclear force.

The *Viking I* probe lands on Mars.

Mao Zedong dies, opening the way for moderate government in China.

pressure on the EPA as well as the chemical, petroleum, and pharmaceutical industries to determine how to measure most effectively these priority pollutants, especially in water discharges. Enter gas chromatography mass spectrometry.

Commercialization of Gas Chromatography Mass Spectrometry

When the EPA was created, gas chromatography (GC) was the most widely used technique for environmental analysis. There was, however, a small number of individuals working in the field of environmental analysis and in the instrumentation industry who were investigating the possibilities of GC-MS as a rival method for environmental analysis. The Finnigan Instruments Corporation, an instrument company established in 1967 to manufacture and sell quadrupole mass spectrometers, was especially interested. Finnigan linked GC to its new quadrupole mass filter technology, and in 1968 the company introduced one of the first commercial gas chromatograph mass spectrometers. Hewlett-Packard also saw a market for GC-MS and followed Finnigan three years later into the market with its own instrument.

Anticipating the expansion of its regulatory role to include the identification and evaluation of specific organic pollutants in water and other environmental matrices, the EPA initiated early efforts to select an analytical system that the Office of Water Quality would mandate. In the early 1970s these efforts were strictly in the research phase. The EPA recognized that GC-MS could perhaps meet the needs of environmental analysis, but it had no idea what the scale of application would be and how many instruments would be required to meet that demand. The handful of companies that built GC-MS instruments was equally uncertain. Firms like Finnigan and Hewlett-Packard were simply looking for any market for their GC-MS technology.

To investigate the potentials of GC-MS, the EPA contracted an expert panel to evaluate various commercial instruments and develop standard methods. Among the mass spectrometers considered by the panel were those built by Varian-MAT, Associated Electrical Industries Ltd. of the United Kingdom, LKB of Sweden, Nuclide Corporation, DuPont, Finnigan, and Hewlett-Packard. Two of the instruments evaluated by the panel were quadrupole mass spectrometers, while the other devices were single- or double-focusing magnetic sector machines. Among the panel's review criteria was the requirement of full scan sensitivity for pesticides ranging from ten nanograms to one microgram, with resolving power from 1,000 to 10,000. Another critical review criterion was whether or not the instrument had a data-handling system suitable for instrument control and data manipulation. Other important considerations included analysis speed, cost, and performance of the gas chromatograph.

The panel recommended the Finnigan quadrupole GC-MS instrument to the EPA for further evaluation, although none of the devices that the panel examined was judged to be entirely satisfactory. Shortly thereafter Finnigan received an order for twenty of its Model 1015 GC-MS instruments—a huge order for the fledgling GC-MS market. Moreover, in

MASS SPECTROMETRY	1976	1977	
	Mixture analysis by tandem mass spectrometry is first demonstrated.	American and Australian researchers introduce the tandem quadrupole mass spectrometer.	Accelerator mass spectrometry is introduced.

HISTORY		1977	
		Recombinant DNA technology is first used to produce insulin.	Smallpox is officially eradicated worldwide.

Environmental Distress

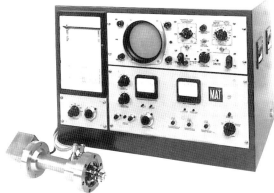

Top, left: Nobel laureate Wolfgang Paul.

Top, right: Brochure for the Finnigan 1015 gas chromatograph mass spectrometer.

Bottom: Atlas-MAT's AMP 3, the first commercial quadrupole residual gas analyzer.

terms of technology, the EPA's purchase of the Model 1015 was a daring move. The quadrupole mass filter was a relatively new mass analyzer in the analytical mass spectrometer market, and it had not yet been tested extensively in the field.

Origins and Evolution of the Quadrupole Mass Spectrometer

The concept behind the quadrupole mass analyzer was introduced in 1953 by physicist Wolfgang Paul. Interestingly, physicist Richard Post suggested in a 1953 Lawrence Radiation Laboratory report that quadrupole technology could possibly serve as an isotope separator for the enrichment of uranium. By the early 1960s several instrument manufacturers in the United States and Europe had commercialized quadrupole technology for residual gas analysis. Atlas-MAT, based in Bremen, Germany, introduced the first residual gas analyzer based on the quadrupole design in 1961, but it did not enjoy widespread distribution. Following Atlas-MAT's lead in the 1960s, the American firms ExtraNuclear, Varian, and Electronics Associates Incorporated (EAI) produced quadrupole-based residual gas analyzers for experimental physicists to use.

EAI was particularly successful in delivering quadrupole gas analyzers to the physics and space science market, selling more than five hundred instruments. EAI exploited this success to underwrite the development of the first quadrupole for use as a commercial analytical mass spectrometer in 1966.

At this time EAI attempted a partnership with Hewlett-Packard (HP) in which HP would resell the EAI quadrupole, bundled with HP gas chromatographs. Dissatisfied with this effort, however, EAI followed with the introduction of a new instrument, the Quad 300D—a quadrupole mass spectrometer with a glass-column gas chromatograph inlet.

The commercial difficulties of EAI's new GC-MS as well as the failure of the company to sell this business to the pharmaceutical firm Syntex frustrated those working on quadrupole technology at EAI. Consequently, EAI employees Robert Finnigan and Michael Story decided to leave the firm and strike out on their own in 1967. Joined by William Fies, an electronics specialist who had worked on quadrupoles at the Stanford Research Institute, Finnigan and Story founded Finnigan Instruments Corporation, with the goal of developing and marketing successful quadrupole GC-MS instruments to bench chemists.

When Finnigan introduced its first quadrupole GC-MS instrument, gas chromatography, which used flame ionization or electron capture detectors, was the technology of choice

1978

The simulated ion optics programme (SIMION) is designed and developed.	A kinetic method to determine gas-phase proton affinities is introduced.	Hydrogen-deuterium exchange as a structural tool for the study of gas-phase ions is developed at the University of Colorado.	Electron-capture negative ion chemical ionization mass spectrometry extends detection limits for derivatized compounds to the attomole level.

1978

Raymond V. Damadian introduces magnetic resonance imaging (MRI) for medical diagnosis.	Egypt and Israel sign the Camp David Accord.	Joy Louise Brown, the first "test-tube baby," is born in England.	Followers of Reverend Jim Jones commit mass suicide in Jonestown, Guyana.

Electronics Associates, Inc., Quad 300, the first commercial quadrupole mass spectrometer.

for the analysis of organic compounds and mixtures. Although widely used, GC alone lacked the specificity that GC-MS could provide.

In 1968, when Finnigan introduced its first quadrupole GC-MS instrument, experts were ambivalent about the quadrupole as a mass analyzer. Magnetic sector instruments dominated analytical mass spectrometry. Universities offered courses in their design and use, and many doctoral dissertations were based on research conducted with them. Researchers used magnetic sectors for their published work in the advancement of analytical chemistry, organic chemistry, physics, biochemistry, and instrument technology itself. Industrial laboratories employed university-trained individuals to operate and interpret the data produced by these large and expensive machines. A large body of literature detailed magnetic sector instruments and their use.

Despite its novelty as an analytical mass spectrometer, the quadrupole mass filter held several advantages over magnetic sector instruments for use as an organic chemist's gas chromatography "detector." The quadrupole instrument was smaller than a magnetic sector instrument and was also less costly. Quadrupole instruments were well suited for use with gas chromatographs in that they could operate more easily at higher pressures than magnetic sectors, crucial when dealing with the gas streams eluting from the chromatograph. Furthermore, the scanning speed of the quadrupole matched closely the speed at which separated compounds emerged from the gas chromatograph. And quadrupole mass spectrometers were more immediately amenable to control by data systems than magnetic sector instruments—they were easier to automate.

Even with these advantages over magnetic sector mass spectrometers, quadrupole GC-MS instruments were by no means easily moved from the drawing board to the laboratory bench top. In 1971, when the EPA's expert panel evaluated available GC-MS instruments, they were still temperamental machines that could easily occupy as much as twenty to forty square feet of floor space. Data system operation was typically performed with the use of punched-paper tape and teletype commands, along with front panel switches. The mass spectral data still had to be read and interpreted manually. The quadrupole analyzers were susceptible to contamination and often required a chemist with an advanced degree and specialized knowledge of mass spectrometry to operate them. Finally, the chemical analysis methods necessary for environmental regulation had not yet been developed, and little was known about the instrumental characteristics that environmentally significant samples would require.

Throughout the 1970s the sensitivity and resolution of quadrupole technology improved significantly. Early quadrupoles suffered from low sensitivity and poor resolution at very high masses—above 500 daltons. Alternative designs and construction parameters were applied to enhance the performance of the quadrupole analyzers. In 1971 Hewlett-Packard introduced a new "dodecapole" mass analyzer built with four rods and eight parallel tuning electrodes. A year later Universal Monitor patented a "monolithic" quadrupole. ExtraNuclear Laboratories

MASS SPECTROMETRY

1979
Researchers in Canada develop the theory of field-induced ion evaporation.

HISTORY

1978
Several hundred Vietnamese refugees drown when their boats sink during the "boat people" crisis.

1979
In Cambodia, invading Vietnamese forces overthrow the Khmer Rouge.

The most serious nuclear accident in U.S. history occurs at Three Mile Island power plant near Harrisburg, Pennsylvania.

Environmental Distress

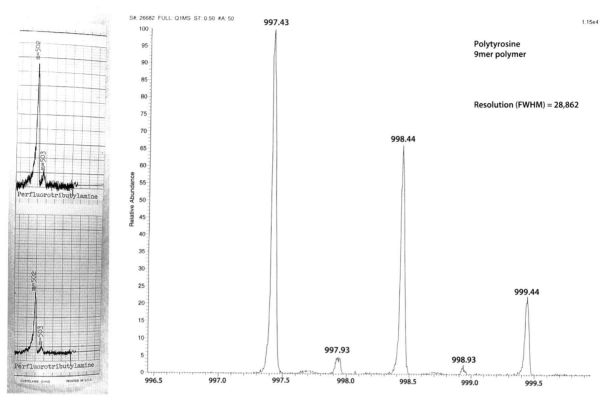

S#: 26682 FULL: Q1MS ST: 0.50 #A: 50

1.15e4

**Polytyrosine
9mer polymer**

Resolution (FWHM) = 28,862

997.43

998.44

999.44

997.93

998.93

Quadrupole mass spectrum of per-fluorotributylamine taken in 1968. The mass resolution is 700.

Right: Mass spectrum of quadruply charged melittin at mass 713 recorded with the Finnigan Quantum, demonstrating the high resolving power of modern quadrupole mass filters.

announced a remarkable increase in mass resolution—as high as 8,000—with the use of larger quadrupole structures. Quadrupoles with hyperbolic rather than round rods were introduced by Varian MAT in 1976 and by Franzen Analysentechnik two years later. The quality of quadrupole performance has continued to improve. By 1980 quadrupole technology was well suited for mass analysis, and its superiority over other instruments for GC detection in environmental analysis was no longer in question. Today, quadrupole mass filters have achieved resolving powers as high as 10,000 with reasonable sensitivity.

A major hurdle that had to be overcome for GC-MS to become a viable commercial instrument centered on interfacing its two primary components. The gas chromatograph and the mass spectrometer operated at different pressures. The outlet of a gas chromatograph typically operated at atmospheric pressure, while the ion source in a mass spectrometer functioned at much lower pressures, ten million times lower for quadrupoles and one hundred million times lower for magnetic sector instruments. To bridge this gap, innovators grappled with both technologies. Glass jet separators were made commercially using Silicon Valley semiconductor technology. Finnigan solved a critical noise problem by moving the ion detector off the main axis of the instrument. These features allowed quadrupole mass

The Sandanistas gain control of Nicaragua.

The Shah of Iran is overthrown, and the Ayatollah Khomeini becomes leader of the country.

The Soviet Union invades Afghanistan.

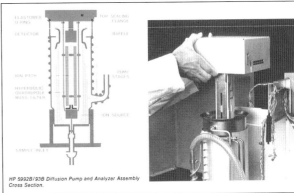

HP 5992B/93B Diffusion Pump and Analyzer Assembly Cross Section.

Top, left: Hewlett-Packard 5970 mass selective detector (MSD). The original MSD, this instrument represented a dramatic decrease in both the size and price of mass spectrometers and was the instrument that "brought mass spectrometry to the masses." Shown here with the Hewlett-Packard 5790 gas chromatograph.

Top, right: Diagram of the Hewlett-Packard 5992 quadrupole analyzer in the diffusion pump. On the right is the actual analyzer assembly.

Bottom: Brochure for the Hewlett-Packard 5992A, the first bench-top GC-MS instrument.

spectrometers to surpass the previous microgram detection limits by several orders of magnitude, a critical advance required for environmental analysis. Most recently, capillary chromatography columns with reduced helium flow were introduced, along with high-speed differentially pumped ion sources, obviating the need for any separator interface.

Improvements in quadrupole technology during the 1970s were matched by a persistent drive toward miniaturization of the instrument itself. Hewlett-Packard played a leading role in this process. In 1971 HP marketed its first quadrupole GC-MS as a general-purpose laboratory instrument. Five years later the company introduced the Model 5992A, the first commercial bench-top GC-MS instrument. Its radical design stressed miniaturization and included placement of the quadrupole mass analyzer inside the diffusion pump. Although there were early skeptics of this design, the Model 5992A set the expectations for the bench-chemist market. Further developments led to the Model 5971, the first complete mass spectrometer smaller than the gas chromatograph itself. This new generation of compact quadrupole GC-MS instruments culminated in the highly successful mass selective detector (MSD), introduced by HP in 1982. Thousands of MSDs were sold, more than any other type of analytical mass spectrometer.

Acquiring and Interpreting the Data

GC-MS instruments had improved significantly by 1980. They were fast, reliable, and small. However, a remaining bottleneck to their usefulness was their limited capacity to handle the flood of data coming from the mass spectrometer. GC-MS instruments generated an enormous amount of data in a very short period. Traditional manual methods of translating the analog recording of a mass spectrum into a list of masses and intensities could not keep pace. Although digital computer technology in the United States was still in its infancy in the early 1960s, its use for data collection, manipulation, and presentation of mass spectral data began to replace the time-consuming, tedious calculations carried out manually by data analysts working with conventional laboratory recording methods.

MASS SPECTROMETRY	**1980**		
	Inductively coupled plasma mass spectrometry is developed at Iowa State University.	GC-MS is used in the investigation of the hazardous waste disposal site at Love Canal.	The first commercial triple-quadrupole mass spectrometers are introduced.

HISTORY	**1980**		
	Lech Walesa helps found the Solidarity labor union in Poland.	The Iran-Iraq war begins.	Mt. St. Helens erupts in Washington State.

Data Systems for Mass Spectrometry

While the development of a data system particularly suited to the demands of GC-MS instruments was critical to the productive use of mass spectrometry in environmental analysis, efforts to computerize the mass spectral data-acquisition process had been under way for a number of years. The earliest attempts to apply automation to the tedious task of reducing the graphic mass spectrum to a tabular listing of mass and intensity was done in the late 1950s when Consolidated Engineering Corporation introduced the Mascot mass spectrum digitizer. This device took as input the signal from the mass spectrometer and, using a combination of analog and digital electronic circuitry, created a digital spectrum of mass and intensity in real time as the spectrum was scanned. Typical scan speeds at the time were on the order of a minute over a mass range of about 120 daltons. A modified ten-key printing adding machine was used to output the data from the Mascot. Initially, the device was used in parallel with the oscillographic recording system, since mass spectroscopists needed assurance that this "black box" was doing its job correctly.

The problem of speeding up digitizer performance for GC-MS applications was addressed in papers presented in instrumentation sessions at the 12th Annual Conference on Mass Spectrometry and Allied Topics in Montreal in 1964. The approach with digitizers was based on dedicated hardwired devices; however, the desire to apply programmable computers to the problem

was seen as a much better solution. While minicomputers were becoming available with just enough resources to do the job, the price of random access memory (RAM) for these machines made them prohibitively expensive with more than the minimum amount of memory. The rule of thumb at the time for estimating the cost of memory was one dollar per bit. Many a programmer worked well into the wee hours of the morning creating highly efficient binary code and then devising ways to swap it in and out of RAM to perform the complex operations required. The alternative, for those with access to mainframe computers, was to record the data in analog form on magnetic tape and transport the tape to the mainframe for processing. Of course, the mainframe had to have an analog-to-digital converter input to extract digital data from the analog tape.

Magnetic sector manufacturers pursued these avenues independently of data system developments in the quadrupole mass filter market. As minicomputers became more powerful and cheap solid-state memory became available, the programming chore became more manageable; data systems thus became standard on all mass spectrometers, regardless of their mass separation principle. Today a vanilla off-the-shelf personal computer has so many orders of magnitude more computing power and RAM than the early minicomputers that it is hard to imagine that anyone would even consider trying to solve the problem with the limited resources available at the time.

A complex organic mixture for environmental analysis may contain hundreds of individual compounds. Identifying the compounds in the mixture might require hundreds of individual spectra, while determining the quantities of each compound might push the number of required spectra into the thousands. The manual performance of such a task was so time consuming as to render it impossible in practice. Two solutions were successfully exploited to overcome this bottleneck in data handling: a computer system that provided instrument control and quantitation of identified analytes and the creation of large spectral libraries of organic compounds for use with automated search algorithms for chemical identification and verification. Widespread use of GC-MS instrumentation to solve the analytical demands of

1981

Fast atom bombardment is introduced by researchers in the United Kingdom.

1981

Father-and-son scientists Luis W. and Walter Alvarez propose that an asteroid impact caused the extinction of the dinosaurs.

Deng Xiaoping becomes the leader of China.

The first cases of AIDS are identified.

environmental assessment and remediation depended on efficient automation of the collection, manipulation, and presentation of data for expert review.

An early attempt to wed the digital computer to GC-MS took place in Cambridge, Massachusetts, in 1967. Researchers at the Massachusetts Institute of Technology introduced the first computer-assisted digital data-acquisition system for a GC-MS instrument. The raw data from the spectrometer was recorded on magnetic tape. Then the tape was read by a mainframe computer. The computer software identified the mass spectral peaks, assigned them mass values and intensities, and printed the resulting spectrum in numerical and graphical form. Operation and control of the mass spectrometer was still accomplished manually. Additional innovations in computer-based control, data acquisition, and interpretation originated elsewhere.

At the Annual Conference on Mass Spectrometry and Allied Topics held in Dallas, Texas, in 1969, researchers from Stanford University unveiled a minicomputer system for data acquisition from and control of a quadrupole mass spectrometer. The Stanford group, whose members formed System Industries, was already using the Finnigan Model 1015. In fact, the new computer system was designed specifically to work with the 1015. The advantages of this merging of technologies was immediately apparent. By digitally jumping from peak to peak, the mass spectrometer could scan through the entire spectrum of the analyte, staying at each mass peak until an adequate ion current was attained. Since the mass to be recorded was set by the computer, it was also possible to monitor only certain ions for quantitation of a known compound—a process known as multiple ion detection (MID). This technique had been developed earlier, but the quadrupole mass spectrometer was especially well suited to its application. Added flexibility using MID could be achieved through a combination of scanning and peak jumping. This computerized data system was the first of its kind to control the mass spectrometer during mass analysis and also record and manipulate the output. System Industries' System 150, as it was known, was sold specifically for quadrupole mass spectrometers, first through Finnigan and then directly to other quadrupole instrument manufacturers.

In addition to automating data handling and control, new computer technologies opened up new possibilities throughout the 1970s for the interpretation of GC-MS spectra. Interest in the use of digital computers to automate identification of spectra produced by new GC-MS instruments was far from idle curiosity. The environmental testing called for by the new regulations and laws would have the most serious of consequences—fines, plant closures, large remediation projects. In experienced hands mass spectra could provide critically important identifications and quantitative measurements. By the mid-1970s GC-MS had evolved into GC-MS-DS—the integration of a data system into the GC-MS to create a new, unified instrument. However, the scale of environmental testing precluded identifications and quantitation by experts. Laboratory technicians would have to process hundreds of samples every day using these automated GC-MS-DS instruments. How would identifi-

MASS SPECTROMETRY	1981		
	ASMS holds a fast atom bombardment workshop.		

HISTORY	1981		
	The first space shuttle, *Columbia*, is launched.	U.S. hostages held in Iran are released.	Sandra Day O'Connor becomes the first woman to serve on the U.S. Supreme Court.

cation and quantification keep pace with this volume of data, and could the data still be trusted? The automation of these tasks with digital computers answered these questions and also addressed the question of cost.

Although the System 150 set new expectations for computerized mass spectrometer data systems, the unique requirements of the environmental marketplace were only realized in the Incos data system, designed originally to control magnetic sector and quadrupole instruments. After Finnigan bought Incos Corporation in 1976, the data system was redesigned for environmental analysis and reporting. The new Incos system, introduced in 1978, became the model for most subsequent systems used for environmental analyses. The key to the automation of compound identification was the creation of a digital library containing the mass spectra of known compounds and the development of computer algorithms that would allow the comparison of spectra from GC-MS-DS instruments with those of the library. In the 1970s the National Institutes of Health (NIH) digitized and expanded a library of mass spectra originally published by the American Petroleum Institute. By the end of the decade responsibility for this digital library was transferred to the EPA.

The EPA was quickly confronted with serious issues surrounding the use of automated identification with the digital library as a critical component in large-scale environmental testing. How could the EPA ensure the fidelity and reliability of these automated identifications? The digital library had been built with mass spectra produced on a variety of different instruments. This raised the issue of how mass spectra could be compared across instruments in a way that would ensure accurate automated identifications. Furthermore, depending on the settings and calibrations of particular kinds of mass spectrometers, variations could be produced in the mass spectra generated from the same compound.

In the late 1970s the EPA developed practices to address these concerns. As part of its mandated methods for environmental analysis, the EPA required that any GC-MS-DS instrument that was going to produce an automated identification using the digital library as part of an EPA environmental test had to meet specific calibration requirements. The instruments had to be able to produce a mass spectrum of a standard compound—decafluorotriphenylphosphine (DFTPP)—within a specific relative abundance range. In addition, commercial producers of mass spectrometers developed and provided special algorithms that allowed the spectra produced on their instruments to be compared with any spectra in the digital library—regardless of their instrument of origin—in the process of automated identification. Currently, the National Institute of Standards and Technology (NIST) oversees this library, which has grown to include more than a hundred thousand spectra.

Selecting Gas Chromatography Mass Spectrometry as the "Method of Choice"

The 1976 consent decree between the EPA and environmental activists on the Water Pollution Control Act Amendments specified 129 priority pollutants, including 13 metals, cyanides, and more than 100 organic chemicals, some of which were industrial mixtures.

1982

GC-MS is included as an approved analytical method in the U.S. Environmental Protection Agency's Resource Conservation and Recovery Act and the Superfund program.

A complete insulin spectrum is obtained by FAB and particle desorption ionization methods.

1982

Surgeon William C. DeVries installs a Jarvik-7 artificial heart in Barney Frank.

Britain defeats Argentina in the Falkland Islands War.

A worldwide ban on commercial whaling is implemented.

HS&T
SPECIAL REPORT

Priority pollutants
II—cost-effective analysis

This list of 129 toxic pollutants includes 114 organic chemicals (with 17 pesticides and 7 PCBs) and 15 inorganic metals or ions. While not yet an approved EPA method, automated gas chromatography/mass spectrometry (GC/MS) is obviously the coming trend

Robert Finnigan and his associates published their comparison of GC-MS and GC alone in the journal Environmental Science and Technology *in May 1979.*

Because of the decree's guidelines the EPA had to establish standard analytical methods to measure these compounds. By the mid-1970s two analytical methods contended for the analysis of all the compounds identified in the consent decree. Government regulators, industrial representatives, and instrument developers were thus drawn into an intense debate over which methods would be mandated—GC alone or GC-MS-DS. GC-MS-DS provided higher chemical specificity and produced fewer false-positive and false-negative results, which made the method more scientifically and legally defensible. Despite this advantage, however, GC-MS-DS involved high capital investment costs.

A pivotal paper comparing the operating costs of GC and GC-MS-DS appeared in 1979. Authored by Robert Finnigan and his company colleagues, the article showed unequivocally that the instrument and labor costs per analyte were higher using stand-alone gas chromatography methods than with a GC-MS-DS instrument. The paper was a key factor in building a consensus in which EPA scientists and outside experts agreed that GC-MS-DS was the preferred analytical method of choice over GC alone. GC-MS-DS was formally proposed as a standard analytical method in the Federal Register in December 1979. The following year the EPA and its regulated industries agreed on GC-MS-DS as the "method of choice" for environmental analysis. The debate finally over, regulators, industry, and analytical instrument companies then moved forward to employ GC-MS-DS on a large scale. Experts would also continue to improve GC-MS-DS, thus providing better tools for understanding and improving the environment.

Applying Gas Chromatography Mass Spectrometry to Environmental Problems

The 1970s and 1980s were marked by high-profile cases of environmental contamination, including the Chernobyl nuclear disaster in 1986 and the massive *Exxon Valdez* oil spill just three years later. Especially notorious was the debacle at Love Canal, in New York State. Located near the city of Niagara Falls, this abandoned canal had served as the disposal site for more than twenty thousand tons of chemical waste products, including hazardous polychlorinated biphenyls, dioxins, and pesticides. When widespread chemical contamination was found near an elementary school and some residential homes, the canal was sealed and the town's inhabitants were moved elsewhere. To deal with problems of this magnitude, Congress passed the Comprehensive Environmental Response, Compensation, and Liability Act in 1980. Also known as "Superfund," this act opened an entirely new field of investigation as literally thousands of sites were identified and examined for the presence of chemicals. Analytical rigor was especially important for Superfund, since it involved the issue of determining liability for contaminated sites. Analytical testing results would be evidence in court.

MASS SPECTROMETRY	1983		
	The theory of charge remote fragmentation is proposed at the University of Nebraska.	The first modern multicollector instrument for continuous-flow isotope ratio mass spectrometry is introduced.	The thermospray interface for liquid chromatography mass spectrometry is announced.

HISTORY	1982		1983	
	Audio compact discs are introduced.		U.S. forces invade Grenada.	British scientist Joe Farman first observes the Antarctic ozone hole.

Environmental contamination at Love Canal in New York State forced local inhabitants to leave their homes and relocate elsewhere.

Since GC-MS-DS was already established as the EPA's method of choice for environmental analysis, it should come as no surprise that it was used at Love Canal and other sites.

Indeed, the Superfund Act created an immediate demand for GC-MS-DS. On the heels of the act the EPA created the Contract Laboratory Program (CLP) to run the millions of samples that would eventually be required to locate and remediate contaminated sites throughout the United States. During the 1980s large environmental testing laboratories—analytical "factories"—were established, requiring many GC-MS-DS systems. The cost of GC-MS-DS instruments continued to fall throughout this period, as did their size in comparison to magnetic sector mass spectrometers. Recently, the EPA Office of Emergency and Remedial Response has observed that "since the inception of the CLP in 1980, over 1,850,000 samples from over 10,000 sites worth approximately $366,000,000 have been analyzed by over 430 laboratories." It is likely that an equal number of samples were run by industrial firms as well. By 1991 some twelve hundred environmental testing laboratories were in operation. All in all, this has probably been the largest analytical effort in history.

In addition to regulating industrial pollutants released into the environment, the EPA is responsible for monitoring the use of chemicals on crops and foodstuffs. The Office of Pesticides Programs, for example, has a large role in determining which chemicals can be used in agriculture. Gas chromatography, and then later GC-MS-DS, was used extensively by this office to monitor compliance. Although DDT is perhaps the most well-known and controversial pesticide, the presence of dioxins in the environment has also drawn considerable attention and scrutiny. Dioxins were originally thought to come only from the production of phenoxy herbicides. However, it has since been determined that they are a common by-product of waste incineration. The compound has a number of isomers, and it was necessary to detect all of them at levels of 10^{-15} grams. Both high-resolution GC and MS were required to analyze dioxins in the environment. Quadrupole mass analyzers were not used because of their limited resolving power. Double-focusing magnetic sector machines were used instead, working at resolving powers as high as 50,000. Early work was done using Kratos MS50 instruments, with later use of Micromass and Finnigan-MAT devices. Because of the high cost to purchase and operate these instruments, however, only a few laboratories performed this type of dioxin analysis.

The EPA also has responsibility for air pollution standards. Air pollution did not receive the early attention paid to water and soil contamination. The Clean Air Act Amendments, passed in 1990, mandated an increase in the number of air pollutants to be monitored from 6 to 189. Among the new compounds on this list are chlorofluorocarbons, also known as CFCs. These synthetic chemicals have been shown to interfere with the protective ozone layer, which filters out harmful ultraviolet radiation in the stratosphere. Although CFC levels have been dropping in recent years, destruction of atmospheric ozone has already been linked to changing global weather patterns and a rising incidence of skin cancer and other melanomas in humans.

1984

The first commercial inductively coupled plasma mass spectrometer is introduced for elemental analysis.	The first commercial ion trap analyzer is introduced.	Yale University researchers use electrospray ionization mass spectrometry to analyze small biomolecules.

1984

Carl Sagan and others warn of the threat of "nuclear winter."	South Korean president Chun Doo Hwan is assassinated by North Korean agents.	Ethiopia is devastated by a famine triggered by civil war.

*President George Bush signing the
Clean Air Act Amendments in 1990.*

GC-MS-DS technology has allowed environmental researchers to investigate issues on a global scale. This technique has generated valuable data on the worldwide transport of certain persistent chlorinated hydrocarbon pesticides and industrial organic compounds, such as polychlorinated biphenyls (PCBs). Such global transport was only recognized and quantified after analytical tools such as GC-MS-DS were introduced, with related sampling programs in the Great Lakes and northern climates, including the arctic regions. These studies eventually showed that semivolatile compounds—dieldrin, DDT, aldrin, heptachlor, the hexachlorocyclohexanes, chlordane, toxaphene, heptachlor epoxide, and mirex—used in the agricultural tropical and temperate regions of the Earth were being transported to and deposited in the colder climates. This phenomenon occurred as a result of repeated vaporization during warm weather, northern transport on prevailing winds, and deposition on cold days. These compounds eventually became part of the permanent ecosystem of arctic and northern Canada, where there is no agriculture. They are found in fish tissue, polar bear tissue, and ice packs and can persist in the environment for decades, often longer. In many ways GC-MS-DS has enabled the creation of macro-environmental models for study.

Government agencies, industries, researchers, and instrument companies have partnered to achieve remarkable results in environmental improvement during the last thirty years. Consider the following statistics. The concentration of PCBs in herring gull eggs from colonies in the Great Lakes fell 84 percent between 1974 and 1996. The tendency of herring gull eggs to accumulate PCBs provided a sensitive indication of the presence of these dangerous chemicals in the environment. Similarly, the concentration of chlorophenyl and DDT compounds in fall-run coho salmon skins taken from Lakes Michigan and Erie fell nearly 50 percent between 1980 and 1988. The inventory of Superfund hazardous waste sites fell from a high of 39,099 in 1994 to 9,245 in 1997. These sites were exhaustively investigated with high-resolution-fused silica capillary column GC-MS-DS instruments and were either remediated if contamination was found or removed from the list if not considered hazardous. Meanwhile the production of pollutants released into the air, water, and soil by all American industries dropped sharply from 3.35 billion pounds in 1988 to 1.82 billion pounds eight years later. Finally, emissions of volatile organic compounds experienced a 30 percent reduction, from 14.3 million tons in 1970 to 9.83 million tons in 1997.

Changing public attitudes, new analytical techniques, and careful monitoring of industrial pollutants have helped mitigate past and current effects on the environment. It was only with sophisticated analytical instrumentation—GC-MS in particular—that these environmental issues could be fully addressed. While the EPA and legislation created an institutional framework for responding to public concerns about the environment, it was the analytical technology of GC-MS that provided the necessary tools for action—for research, legislation, litigation, reform, and remediation.

MASS SPECTROMETRY	1984		
	Liquid chromatography is interfaced to mass spectrometry with pneumatically assisted electrospray.	A particle beam interface for liquid chromatography mass spectrometry is developed at the Georgia Institute of Technology.	The Asilomar Conference on Mass Spectrometry affiliates with the ASMS.

HISTORY	1984		
	HIV is identified as the cause of AIDS.	Alec Jeffreys develops genetic fingerprinting.	Hundreds of Sikh fundamentalists are killed in a clash with the Indian army in the city of Amritsar.

Suggested Reading

D. Barcelo et al. "New Developments in Environmental Mass Spectrometry." *Advances in Mass Spectrometry* 13 (1995), 465–478.

C. Brunnée; L. Delgmann; E. Kronenberger. "Construction Details and Performance of a New Commercial Quadrupole Filter." In *Proceedings of the 11th Annual Conference on Mass Spectrometry and Allied Topics*, 388–397. Philadelphia: American Society for Testing and Materials, 1963.

W. L. Budde. *Analytical Mass Spectrometry: Strategies for Environmental and Related Applications.* New York: Oxford University Press; Washington, D.C.: American Chemical Society, 2001.

R. Carson. *Silent Spring.* Boston: Houghton Mifflin, 1962.

C. G. Daughton. "Emerging Pollutants, and Communicating the Science of Environmental Chemistry and Mass Spectrometry: Pharmaceuticals and the Environment." *Journal of the American Society for Mass Spectrometry* 12 (2001), 1067–1076.

J. W. Eichelberger et al. "Reference Compound to Calibrate Ion Abundance Measurements in Gas Chromatography–Mass Spectrometry Systems." *Analytical Chemistry* 47 (1975), 995–1000.

Environmental Quality. 23rd through 27th Annual Reports (1993–1997). Washington, D.C.: Council on Environmental Quality, Office of the President, U.S. Government Printing Office.

B. Erikson. "Environmental Protection Agency." *Today's Chemist at Work* 8 (1999), 101–124.

L. S. Ettre. "The Development of Gas Chromatography." *Journal of Chromatography* 112 (1975), 1–26.

R. E. Finnigan. "Quadrupole Mass Spectrometers: From Development to Commercialization." *Analytical Chemistry* 66 (1994), 969A–975A.

R. E. Finnigan; D. W. Hoyt; D. E. Smith. "Priority Pollutants II: Cost Effective Analysis." *Environmental Science and Technology* 13 (1979), 534–541.

B. K. Gullett et al. "Emissions of PCDD/F from Uncontrolled, Domestic Waste Burning." *Chemosphere* 43 (2001), 721–725.

H. S. Hertz; R. A. Hites; K. Biemann. "Identification of Mass Spectra by Computer Searching a File of Known Spectra." *Analytical Chemistry* 43 (1971), 681–691.

R. A. Hites; K. Biemann. "Mass Spectrometer-Computer System Particularly Suited for Gas Chromatography of Complex Mixtures." *Analytical Chemistry* 40 (1968), 1217–1221.

R. Jaffe and R. A. Hites. "Environmental Impact of Two, Adjacent, Hazardous Waste Disposal Sites in the Niagara River [USA] Watershed." *Journal of Great Lakes Research* 10 (1984), 440–448.

Maynard B. Neher. *Summary Report on Evaluation of Gas Chromatograph/Mass Spectrometer/Computer Systems to Water Quality Office, Environmental Protection Agency.* Columbus, Ohio: Battelle Columbus Laboratories, 1 June 1971.

W. E. Reynolds et al. "A Computer Operated Mass Spectrometer System." *Analytical Chemistry* 42 (1970), 1122–1129.

Direct Quotation

Page 117, Erikson, 1999, p. 109.

"Absorbing" compounds are used to assist laser ionization of "nonabsorbing" compounds.	A. O. Nier publishes his design of a miniature, field-portable Mattauch-Herzog geometry mass spectrometer.	Researchers at Rice University discover stable C_{60} clusters using time-of-flight mass spectrometry.

Mikhail Gorbachev becomes leader of the Soviet Union.	Mexico City is hit by a major earthquake that kills tens of thousands.	The United States imposes economic sanctions on South Africa.

Chapter **9**

Law and Order

All analytical outcomes are important. It takes the utmost care to ensure a true and accurate result. However, the results of analysis in the forensic sciences may have criminal and civil consequences. Because mass spectrometry, particularly when combined with gas or liquid chromatography, can provide absolute, unambiguous identification of molecules, it is one of the most powerful tools available to forensic scientists. This powerful feature, however, comes with a disadvantage. Mass spectrometry is a destructive technique. Once the analysis is performed, the sample is gone. This disadvantage is balanced by the fact that extremely small amounts of sample, on the order of milligrams or less, are sufficient for analysis. This chapter reviews the history of mass spectrometry in the forensic sciences and highlights some of its more common applications.

The Growth of Forensic Science in the Early Twentieth Century

Scientific crime detection in the United States can be traced back to the 1920s, when the first crime laboratories were established in Los Angeles and Chicago. In the early 1930s the Federal Bureau of Investigation (FBI) established forensic laboratories in New York City and Washington, D.C. Professional training in this field emerged at a select group of academic institutions during this period. The University of California at Berkeley, the City University of New York, and Harvard, Michigan State, and Northwestern Universities all offered instruction in the application of scientific techniques to criminal investigation. By

*Mass spectrometry helps maintain the integrity and competitiveness of athletic competitions, including Olympic events. Seoul 1988: The Start.
©Allsport.*

MASS SPECTROMETRY | **1986**

Liquid chromatography is interfaced to mass spectrometry with pneumatically assisted electrospray.

HISTORY | **1986**

Corazon Aquino is elected president of the Philippines after dictator Ferdinand Marcos is forced to flee the country.

The Iran-Contra scandal is revealed.

the early 1980s nearly 250 public crime laboratories employing approximately 3,500 scientists were operating throughout the United States. And in Europe, where forensic science originated during the late nineteenth century, the number of laboratories and practitioners had also grown substantially. In England and Wales, for example, more than 600 scientists were working in nine government crime laboratories during the early 1980s.

By its very nature forensic science is interdisciplinary, often drawing broadly on research in chemistry, physics, and biology. It is also dependent on expertise in such specialized fields as microscopy, microchemistry, optical spectroscopy, serology, immunology, and crime scene reconstruction. In the 1970s gas chromatography mass spectrometry (GC-MS) joined the arsenal of analytical weapons that forensic chemists could use in criminal investigation.

Much of the science behind forensic mass spectrometry was well known long before the technique was actually used in the laboratory. Consider, for example, the analysis of accelerants used in arson cases. In most cases accelerants are composed of mixtures of hydrocarbons, such as gasoline or kerosene. As we saw in chapter 3, analysis of these compounds and knowledge of their spectra was an advanced field of study by the late 1940s, largely because mass spectrometry had been exploited by researchers in the chemical and petroleum industries. Application of mass spectrometry to arson investigations, however, was delayed until the early 1970s owing to the difficulties of handling extremely complex mixtures. A suitable solution to this problem appeared in the form of GC-MS, which was capable of analyzing the complex mixtures encountered in arson investigations and was quickly adopted by researchers in forensic science laboratories. Especially appealing was the increased specificity and low cost of GC-MS instrumentation. Just as we saw in the chapter on environmental science, forensic science was a beneficiary of the ongoing drive among manufacturers to design, build, and market more powerful and less costly mass spectrometers.

More recently, increased federal spending for crime mitigation has facilitated the adoption of mass spectrometry in forensic laboratories. This transformation is reflected in the number of forensic publications in the scientific literature that have used mass spectrometry over the last several decades (Figure 1). It should be kept in mind, however, that older analytical methods, such as differential thermal analysis, ultraviolet and infrared spectrophotometry, and X-ray fluorescence, continue to find widespread use in forensic laboratories.

In the 1990s forensic science got an additional boost when liquid chromatography mass spectrometry (LC-MS) became more routine. Forensic scientists often deal with molecules extracted from biological fluids or tissue samples, both complex and dirty matrices. Liquid chromatographic techniques are much more amenable to these types of biological samples than is gas chromatography. The recent development of electrospray ionization and efficient interfaces for LC-MS has been a boon to the use of this technique in forensic laboratories.

During criminal investigations forensic scientists are often faced with the difficulty of analyzing a wide variety of substances. Hydrocarbons are studied in arson cases, steroids in athletic and sporting events, and drugs of abuse in criminal investigations. In order to ana-

MASS SPECTROMETRY

1987
Capillary electrophoresis mass spectrometry that uses electrospray ionization is introduced.

HISTORY

1986
A nuclear reactor in Chernobyl in the U.S.S.R. melts down and explodes in the worst nuclear power accident in history.

1987
Nazi war criminal Klaus Barbie is sentenced by a French court to life in prison.

Figure 1. Publications of forensic applications of mass spectrometry.

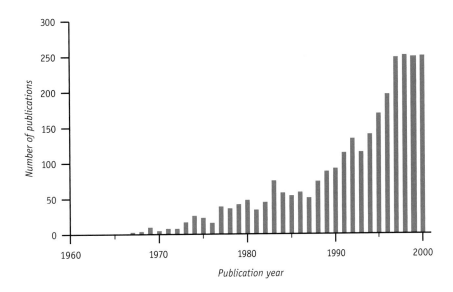

lyze these classes of materials, different experimental approaches are required. The forensic scientist is also obliged to interpret these data rigorously and accurately for the benefit of due process in the legal system.

This work carries with it several formidable tasks when mass spectrometric data is to be entered into evidence, not the least of which is to communicate the findings of the analysis to a jury whose members may be scientifically naïve. Testimony may include a description of the operational principles of a mass spectrometer and the data it generates. It would be futile for either side of a court battle to rely on the dense technical jargon that normally pervades the mass spectrometry community.

Mass Spectrometry and Performance-Enhancing Drugs

One of the first publications to describe mass spectral analysis of performance-enhancing drugs appeared in the *Journal of Pharmacy and Pharmacology* in 1967. In this article researchers at the University of London identified GC-MS as the best technique to detect performance-enhancing drugs in professional athletes. They also recommended that the administrative oversight of international sporting events include provisions to monitor the use of such drugs by participating athletes. GC-MS was identified as the critical analytical tool for such an undertaking.

Their recommendation was put in place for the 1968 Winter Olympics in the French city of Grenoble. Only a few stimulants and narcotics were checked, partly because some of the more sophisticated designer drugs had not come into widespread use. Most of the drugs

1988

Electrospray ionization of proteins with molecular weights in the range of 5,000 to 40,000 daltons is demonstrated.

1988

The United States and the U.S.S.R. agree to arms reduction in the Intermediate-Range Nuclear Forces Treaty.

The Nicaraguan civil war ends.

Soviet forces begin to withdraw from Afghanistan.

were well within the analytical capabilities of the instruments used at the time. As the performance of mass spectrometric instrumentation improved, more comprehensive testing became possible. All participating athletes in the 1984 Olympic Games in Los Angeles, for example, were tested for anabolic steroids and their metabolites.

Experiments designed to detect drugs begin with a biological sample, typically urine, but in some cases blood, plasma, or even tissue are collected for analysis. More recently, hair, sweat, and saliva have been used for this purpose. Regardless of its nature and origins, the sample represents a complex matrix from which the target analyte must be isolated and prepared for analysis. The selection of a sample workup protocol also depends on the class of compounds suspected and the type of analysis to be performed. A drug screen analysis is designed to detect the presence of a wide range of possible drugs and can be conducted by a number of methods, such as by gas chromatography alone, radioimmunoassay, or atomic absorption spectroscopy. If the screening of a sample indicates the possible presence of a drug, then a separate confirmational analysis is performed by GC-MS. In some cases the latter analysis may make use of a different sample workup protocol than the one used in the drug screen. This double analysis procedure, while lengthy and time consuming, also helps to guard against false-positive results. Recent developments in tandem mass spectrometry permit these types of analyses to be done with less stringent sample workup protocols.

The classes of drugs most frequently used in competitive sports are stimulants, narcotics, anabolic agents, diuretics, corticosteroids, and beta-blockers. Using mass spectrometry to ensure that competitive athletes are drug free is a constant battle between forensic scientists and the developers of performance-enhancing drugs. As protocols for analysis are developed for known drugs, newer drugs are introduced that may not be detected by existing methods or that are more difficult to isolate. This inability to screen for all drugs puts continuous pressure on forensic chemists and their analytical instruments. New protocols must be devised and implemented. For example, current practice includes the use of high-resolution gas chromatography–high-resolution mass spectrometry for increased specificity in testing of Olympic athletes for performance-enhancing drugs.

Today drug detection at international competitions requires the construction of a temporary laboratory on site that is equipped with analytical instruments and research personnel working around the clock. The sensitivity of the combined GC-MS technique is sufficiently high that evidence of the use of banned substances can be detected up to four days after last use. In the case of marijuana and some other drugs, metabolites can be detected in urine for weeks.

Mass Spectrometry and Horse Racing

While testing for performance-enhancing drugs in competitive sports has been under way for more than thirty years, similar analytical measures have been taken in non-human sports, such as horse racing. The betting public must be assured that the outcome of a horse race is

MASS SPECTROMETRY		1989
		The first ASMS Sanibel Conference takes place.

HISTORY	1988		1989
	Pan Am flight 103 is destroyed by Libyan terrorists over Lockerbie, Scotland.	Warfare erupts between the Soviet Republics of Azerbaijan and Armenia.	Pro-democracy demonstrations in Tiananmen Square are violently quelled by the Chinese military.

Gas chromatography mass spectrometry is now used for drug screening of thoroughbreds at horse-racing events.

not tainted by the influence of performance-enhancing drugs. In 1972 researchers in Japan outlined a procedure whereby GC-MS could be used to identify illegal drugs and their metabolites in horses. It was not until a decade later, however, that the technique was routinely applied to drug screening at horse-racing events. Subsequent improvements by researchers in the United States demonstrated that GC-MS could detect phenothiazine tranquilizers in horse urine at the level of a billionth of a gram. Accuracy of quantification was an especially welcome result, since conventional wet chemical methods could not be used to detect such minute levels of the drug.

Mass Spectrometry, Drugs of Abuse, and Poisons

The use of mass spectrometric techniques to identify drugs of abuse is one of the more frequent applications in forensic science. In many cases the outcome can spell serious consequences for the suspect. Great care must be taken in sample handling, preparation, analysis, and interpretation. The drugs of interest—cannabinoids, marijuana, cocaine and its metabolites, amphetamines, opiates, barbiturates, phenylcyclidine, lysergic acid diethylamide (LSD), benzodiazepines, and fentanyl and its analogs—typically represent a small set of the overall group of drug compounds, which makes the qualitative nature of the analytical problem somewhat simpler. This feature allows the investigator to individualize the analysis to

Wolfgang Paul shares the Nobel Prize in physics for his work on the development of the ion trap.

The first commercial electrospray mass spectrometer is introduced.

The Communist party is voted out of power in Poland's first free elections since before World War II.

The oil tanker *Exxon Valdez* runs aground in Alaska, causing massive damage to coastal ecosystems.

The Berlin Wall is torn down.

target the drug in question. However, such analyses are complicated by the need to detect nanograms per milliliter of the drug or its metabolites in the presence of tens of micrograms of biological background and interfering substances, such as ephedrin and natural hormones. Furthermore, some groups of drugs, such as amphetamines, benzodiazepines, and barbiturates, have very similar electron ionization spectra.

In addition to the presence of the opiate narcotic cocaine, the sample may contain commonly used cutting agents, such as caffeine, sugar, lactose, quinine, lidocaine, and procaine. GC-MS can identify these constituents, thereby providing a clear picture of the total composition of the substance. Drug enforce-

Law-enforcement agencies use mass spectrometry to detect illicit drugs and trace their origins.

French Artillery and Forensic Science

The same techniques used for drug screening are also used for identification of toxic substances that have been ingested, either accidentally or on purpose. An unusual case was reported by researchers in France in 1997. A young French artilleryman had indulged in the tradition of drinking a mixture of beer and wine after it had been used to rinse a hot artillery gun barrel. Shortly afterward he became comatose. Initially, a high blood alcohol level was suspected, but when that test was negative, other more general analytical techniques were enlisted. Inductively coupled plasma (ICP) emission spectroscopy discovered excessively high concentrations of tungsten in the soldier's body fluids, a result confirmed by the recently developed technique of ICP coupled to mass spectrometry. On the basis of these results a fast, reliable quantitative assay by ICP-MS was developed for tungsten and used to follow the effects of various treatments in an attempt to flush the tungsten from the patient's body. The soldier eventually recovered, and the source of the tungsten was found to be from a new alloy used in artillery gun barrels. ICP-MS is fast becoming a powerful analytical tool in forensic and toxicology research.

References

P. Marquet et al. "Tungsten Determination in Biological Fluids, Hair and Nails by Plasma Emission Spectrometry in a Case of Severe Acute Intoxication in Man." *Journal of Forensic Science* 42 (1997), 527–530.

M. A. Chaudhri, R. J. Watling, A. Young. "The Application and Potential of Inductively Coupled Plasma Mass Spectrometry, ICP-MS for High Sensitivity Multi-element Analysis of Medical Samples down to the Sub-PPB Range." In *Trace and Toxic Elements in Nutrition and Health: Proceedings of the 4th International Conference on the Health and Disease: Effects of Essential and Toxic Trace Elements, New Delhi, 8–12 February, 1993,* edited by M. Abdulla, S. B. Vohora, and M. Athar, 421–430. New Delhi: Wiley Eastern, 1995.

MASS SPECTROMETRY	1990		
	The ASMS launches the *Journal of the American Society for Mass Spectrometry*.	The first ASMS Award for a Distinguished Contribution to Mass Spectrometry is awarded.	

HISTORY	1990		
	Namibia gains independence from South Africa.	Civil war breaks out in Liberia.	Iraq invades Kuwait.

ment agencies can use this forensic information to trace illicit drug samples back to their sources. Patterns of drug abuse change over time, and screens and tests for new drugs must be developed to deal with such cases.

While it is easier to obtain evidence of drug use by examination of blood and urine, analysis of body tissues from different organs is sometimes undertaken in postmortem identifications. In each case the careful selection of the best sample workup protocols for the sample in question is an important first step in the analytical procedure. Moreover, quantitative methods must be performed using internal standards, particularly since the analysis protocol frequently requires such steps as extraction and derivatization. The most common standards are isotopically labeled analogs of the drug, either with deuterium or carbon-13. Such analogs share almost identical physicochemical properties with the target drug and are rarely differentiated from the target during sample preparation but are easily distinguished by mass spectrometric analysis, thus making excellent quantitative standards.

Besides drugs of abuse, analysis of biological specimens for poisons represents another important application of combined GC-MS in forensic laboratories. These problems are somewhat more difficult to handle because the range of poisonous substances is so large and, contrary to the situation with drugs of abuse, there may not be any evidence suggesting the

Gas Chromatography Mass Spectrometry Exonerates Mother in Death of Child

In the early 1990s first-time parents in a suburb of St. Louis rushed their sick newborn to the hospital. Within days a double tragedy struck the young couple; their child died and the mother was accused of his death. The murder charge stemmed from a clinical laboratory analysis performed by gas chromatography, which suggested ethylene glycol in the blood serum of the infant. Ethylene glycol tastes sweet and could be easily fed to an infant by someone wishing to harm the child. Since ethylene glycol is the primary ingredient in radiator antifreeze, it would also be readily available to anyone.

While the mother fought charges of murder, she became pregnant a second time and gave birth while awaiting trial. Within weeks this second child, who had been placed in foster care, began to show the same symptoms as his deceased sibling. Because of this occurrence the medical team requested a more thorough clinical laboratory analysis using gas chromatography mass spectrometry rather than gas chromatography alone.

Re-examination of the serum from the child thought to have died of ethylene glycol poisoning showed that it had a rare inborn error of metabolism, methylmalonic acidemia. This condition produces propionic acid as a metabolite in blood serum. Propionic acid elutes from the gas chromatograph at the same time as ethylene glycol. Thus, the gas chromatographic peak identified as ethylene glycol by the clinical laboratory was actually propionic acid. The specificity of mass spectrometry was able to differentiate between the two compounds even though they have the same retention time in gas chromatography. Proof of a metabolic basis for the child's symptoms eventually exonerated his mother of the murder charge.

1991

Oligonucleotides are sequenced using electrospray ionization mass spectrometry.

Cornell researchers investigate noncovalent receptor ligand complexes using mass spectrometry.

1991

Gene therapy is first used to treat human patients.

The Soviet Union dissolves after a failed coup by hard-line Communists trying to reverse the reforms of Mikhail Gorbachev.

The Persian Gulf War is fought between Iraq and the United States.

Arson investigators use mass spectrometry to identify dangerous accelerants.

presence of a particular compound. Such a case can become a challenging analytical hunt involving more than one form of mass spectrometry as well as other analytical techniques.

Mass Spectrometry, Arson, and Explosives

One of the most useful applications of mass spectrometry in the forensic sciences is in the investigation of arson. Accelerants for arson are typically common fuels that are readily available and can be easily purchased by anyone without suspicion. For the purposes of forensic science arson accelerants are broken down into such broad categories as gasoline, medium petroleum distillates, kerosene, heavy petroleum distillates, lamp oils, and turpentine. All of these accelerants are composed of hydrocarbons with a range of boiling points, which, when analyzed by GC-MS, reveal a "fingerprint" characteristic of the fuel. Most individuals with criminal intent assume that the fuel used as an accelerant will be consumed by the fire and thus the evidence of arson will be destroyed. However, in most cases, minute traces of the accelerant remain, and since GC-MS is such a sensitive technique, evidence of arson can still be detected. At the FBI, for example, the detection limit is one tenth of a microliter of gasoline in a four-liter paint can filled with debris.

Typically, the lower boiling components of the accelerant are reduced relative to the higher boiling components during a fire. Standard "fingerprints" are generated by recording both chromatograms and mass spectra of the various hydrocarbons in each potential accelerant before and after having been used in a controlled burn. The presence of a certain pattern of chromatographic peaks and compound types provides unambiguous identification of the accelerant. For instance, kerosene is characterized by a distribution of linear saturated hydrocarbons containing eight to seventeen carbons and clusters of branched hydrocarbons that appear between consecutive pairs of the linear hydrocarbons. Similarly, for each of the categories of potential accelerants listed above, certain characteristic patterns and compounds can be determined.

While the chromatographic pattern is frequently sufficient to indicate the presence of an accelerant from one of the categories above, it is most desirable from a legal perspective to have an unambiguous determination of the various compounds associated with the accelerant. Hence, the ease with which the mass spectrometric data from a GC-MS analysis can be evaluated permits the forensic chemist to plot the intensity of the masses of one or more of the compound types known to be characteristic of the accelerant. For instance, a plot of the intensity of the ion at mass 71 from the file of GC-MS data can readily reveal the pattern of compounds characteristic of kerosene, since mass 71 is a prominent ion in the mass spectrum of saturated hydrocarbons. The application of GC-MS in arson investigations produces data that is nearly incontestable in a court of law owing to the highly specific and detailed information provided by the analysis.

The detection of explosives is another forensic application that depends on the extreme sensitivity of mass spectrometry. Obviously, a crime scene in which explosives are suspected

MASS SPECTROMETRY	1992	
	Researchers report the use of highly enriched ^{13}C-labeled fatty acids to improve the detection limits of high-precision isotope ratio mass spectrometry analysis.	Structural information about compounds is obtained using Reflectron time-of-flight mass spectrometry (MALDI post-source decay).

HISTORY	1991	1992	
	The disintegration of Yugoslavia begins a series of Balkan wars marked by genocide and ethnic hate.	Civil chaos in Somalia prompts a failed U.S. intervention.	Riots sweep through poorer African-American neighborhoods of Los Angeles after the acquittal of police officers who were videotaped beating Rodney King.

Important clues to the causes of major disasters, such as the explosion of TWA Flight 800 in 1996, can be detected using mass spectrometry.

to play a part is not going to reveal much explosive residue, but sufficient traces will remain for detection by mass spectrometric techniques. GC-MS has been the method of choice for explosives investigations, but liquid chromatography mass spectrometry has recently supplanted that analytical tool.

Explosives and their residues have several chemical characteristics that make detection and analysis especially challenging. In particular, the most common ionization technique, electron ionization (EI), is not very useful since the compounds that make up an explosive are by their nature labile; thus they fragment easily under EI conditions. Consequently, the molecular ion for an explosive compound will probably not be observed under such conditions. Therefore investigators have turned to other ionization techniques, such as chemical, thermospray, or electrospray ionization—all techniques that are gentler and produce either a molecular ion or an adduct of the molecular ion with little or no fragmentation. It is much easier to convince a jury that an adduct of the explosive compound is conclusive proof of its presence than it is to claim the same proof from the presence of fragment ions.

Negative ion chemical ionization (NICI) has advantages for the detection of nitro- and nitramine compounds, some of the most common explosives in use. These compounds have a high electron affinity and are thus readily ionized by NICI. An added advantage is that this ionization technique is specific for these explosive compounds, thus they can be detected in fairly complex matrices since few of the matrix components will ionize by NICI.

1993

Rudolph Marcus receives the Nobel Prize in chemistry for his theory of electron transfer reactions in chemical systems.	Structural biologists at Purdue University develop a new tool—protein amide deuterium exchange by mass spectrometry.	Mass spectrometry is applied in the emerging field of "proteomics."

1993

Alberto Fujimori becomes dictator of Peru.	Massive flooding of the upper Mississippi River causes billions of dollars in damage in the midwestern United States.	Eritrea declares its independence from Ethiopia after a long civil war.

In recent years the use of electrospray ionization with liquid chromatography mass spectrometry (LC-ESI-MS) has become more prevalent in forensic science. Since liquid chromatography is a room-temperature process, it is advantageous in the analysis of labile explosive compounds that cannot survive the temperatures encountered during a gas chromatographic analysis. The development of thermospray and electrospray ionization techniques has made the coupling of a liquid chromatograph to the mass spectrometer a practical reality instead of a tool reserved primarily for use in research laboratories. Here again negative ion spectra can be obtained from much less sample than in positive ion mode. Today, LC-ESI-MS is the method of choice for trace explosives analysis in national crime laboratories worldwide. Interestingly, electrospray ionization, an ionization technique successfully applied to biological studies, is also widely used in forensic science for a completely different class of compounds.

The majority of investigations of explosives are after the fact; that is, they attempt to ascertain the type and possible origins of the explosive from analysis of post-blast residues. A desirable analytical goal in this area of forensic science would be to devise a method of screening items for the presence of explosives before the fact—before they are used in an illegal or terrorist activity. In the 1990s FBI researchers reviewed the requirements for such an application and concluded that at the time mass spectrometry did not have the speed and sensitivity to screen luggage for explosives at the rate required—ten items per minute. However, recent developments in a related field, ion mobility spectrometry (IMS), are currently being applied in the scrutiny of selected carry-on items at airports today. In the near future a portal for screening of individuals could be introduced.

Clearly mass spectrometry has been a valuable tool in the forensic sciences. Thirty years ago forensic scientists depended on their own experience and limited experimental tools to help stem the tide of illegal activities, ranging from illicit drug use to arson. In the years that followed, forensic scientists used the analytical power of mass spectrometry to bring precision and accuracy to their investigative skills. Today mass spectrometry is a valued tool in the arsenal of analytical techniques used in the fight against crime. And intensive research is currently under way to create miniaturized mass spectrometers—the size of a softball—which may find widespread application in the forensic sciences in the future.

Suggested Reading

A. H. Beckett; G. T. Tucker; A. C. Moffat. "Routine Detection and Identification in Urine of Stimulants and Other Drugs, Some of Which May Be Used to Modify Performance in Sport." *Journal of Pharmacy and Pharmacology* 19 (1967), 273–294.

W. Bertsch; G. Holzer. "Analysis of Accelerants in Fire Debris by Gas Chromatography/Mass Spectrometry." In *Forensic Applications of Mass Spectrometry,* edited by J. Yinon, 129–169. Boca Raton, Fla.: CRC Press, 1995.

C. E. Costello et al. "The Routine Use of a Flexible GC-MS-Computer System for the Identification of Drugs and Their Metabolites in the Body Fluids of Overdose Victims." *Clinical Chemistry* 20 (1974), 255–265.

MASS SPECTROMETRY	**1994**
	Microelectrospray and nanoelectrospray ionization are introduced.

HISTORY	**1993**	**1994**		
	Andrew Wiles solves Fermat's last theorem.	Ethnic massacres in Rwanda leave nearly 500,000 people dead.		Nelson Mandela is elected president in South Africa's first post-apartheid elections.

J. R. Ehleringer et al. "Tracing the Geographical Origin of Cocaine." *Nature* 408 (2000), 311–312.

D. D. Fetterolf. "Detection and Identification of Explosives by Mass Spectrometry." In *Forensic Applications of Mass Spectrometry,* edited by J. Yinon, 45–57. Boca Raton, Fla.: CRC Press, 1995.

S. M. Gerber. *Chemistry and Crime: From Sherlock Holmes to Today's Courtroom.* Washington, D.C.: American Chemical Society, 1983.

S. M. Gerber; R. Saferstein, eds. *More Chemistry and Crime: From Marsh Arsenic Test to DNA Profile.* Washington, D.C.: American Chemical Society, 1997.

J. D. Henion. "Direct Injection Micro-Liquid Chromatography/Mass Spectrometry Applied to Equine Drug Testing." In *Proceedings of the 3rd International Symposium on Equine Medication Control,* 133–140. Lexington, Ky.: International Equine Medication Control Group and Department of Veterinary Science, College of Agriculture, University of Kentucky, 1980.

P. Kintz. "Hair Testing and Doping Control in Sport." *Toxicology Letters* 102–103 (1998), 109–113.

T. Matsumoto et al. "Detection and Identification of Amphetamine, Methamphetamine, Ephedrine, and Methylephedrine in Horse Urine by GC and GC-MS." In *Proceedings of the 3rd International Symposium on Equine Medication Control,* 77–88. Lexington, Ky.: International Equine Medication Control Group and Department of Veterinary Science, College of Agriculture, University of Kentucky, 1980.

C. K. Meng; M. Mann; J. B. Fenn. "Of Protons or Proteins." *Zeitschrift für Physik. D—Atoms, Molecules and Clusters* 10 (1988), 361–368.

G. R. Nakamura et al. "Forensic Identification of Heroin in Illicit Preparations Using Integrated Gas Chromatography and Mass Spectrometry." *Analytical Chemistry* 44 (1972), 408–410.

J. Park et al. "Mass Spectrometry in Sports Testing." In *Forensic Applications of Mass Spectrometry,* edited by J. Yinon, 95–128. Boca Raton, Fla.: CRC Press, 1995.

M. L. Vestal. "Studies of Ionization Mechanisms Involved in Thermospray LC-MS." *International Journal of Mass Spectrometry and Ion Physics* 46 (1983), 193–196.

M. Yamashita; J. B. Fenn. "Application of Electrospray Mass Spectrometry in Medicine and Biochemistry." *Iyo Masu Kenkyukai Koenshu* 9 (1984), 203–206.

J. A. Zoro; K. J. Hadley. "Organic Mass Spectrometry in Forensic Science." *Journal of the Forensic Science Society* 16 (1976), 103–114.

1995

| Comet Shoemaker-Levy 9 crashes into Jupiter. | The North American Free Trade Agreement joins the economies of Canada, the United States, and Mexico. | An earthquake hits Kobe, Japan, killing over 5,000 people. |

Chapter **10**

*The Community of
Mass Spectrometrists*

Early mass spectrometers, like those introduced by Arthur Dempster, Walker Bleakney, and Alfred Nier, were unique laboratory instruments often crafted from handmade components. They were not easily reproduced. Nor were these early instruments exploited by a large community of users. It was not until World War II that mass spectrometry expanded beyond its academic roots into the world of industrial application. From initial analytical applications in the petroleum industry, mass spectrometry grew rapidly during the postwar years. Environmental science, biochemistry, geochronology and space research, medicine, and forensic science are just a few of the many fields that exploited the analytical capabilities of this technology after 1945. From the first users' group meetings sponsored by Consolidated Engineering Corporation (CEC) in the 1940s to the comprehensive national meetings held today, this transformation has been both dynamic in scope and diverse in content.

Detail from attendees at the National Bureau of Standards meeting on mass spectrometry, 1951. First row from left: Walker Bleakney, Norman Coggeshall, David Hess, Alfred Nier, Kenneth Bainbridge, Edward Condon, Josef Mattuch, and John Hipple. Wolfgang Paul stands immediately behind Hipple, and Harold Washburn is in third row, centered between Condon and Mattauch.

Note: Most of the source material for this chapter was extracted from C. M. Judson, "ASMS Retrospective Notes" and "CEC Group Meetings" (unpublished manuscripts), both preserved at the Chemical Heritage Foundation, Philadelphia.

MASS SPECTROMETRY	1995
	The mass spectrometer on NASA's Galileo probe provides new knowledge about Jupiter's atmosphere.

HISTORY	1995		1996
	The federal building in Oklahoma City is destroyed by a terrorist bomb.	Israeli president Yitzhak Rabin is assassinated.	The Taliban militia gains control of Afghanistan.

From Industry Group Meetings to Committee E-14

The American Society for Mass Spectrometry (ASMS) was founded in 1969. Its origins, however, can be traced back to Southern California during World War II. In February 1944 CEC held the first of its users' meetings at company headquarters in Pasadena. These meetings were designed as information exchanges between CEC personnel and users of the new 21-101 analytical mass spectrometer. Presentations given at these meetings were compiled into reports and distributed to all CEC's customers. Some of the reports were published as articles in the leading scientific and technical journals. The early meetings were typically dominated by researchers from the petroleum and chemical industries and a few government laboratories, such as the National Bureau of Standards and various regional laboratories operated by the Bureau of Mines.

From a small group of only ten participants in February 1944, attendance at the CEC users' meetings rose sharply to thirty-seven by the end of 1945. In 1949 nearly 150 attendees participated in the meeting held in New York City. In the same year General Electric (GE) began sponsoring similar meetings for users of its recently introduced magnetic sector mass spectrometer. By the early 1950s other manufacturers were moving into the commercial instrument business and wanted access to the same type of information exchange that had originated with the CEC meetings. CEC established a strict patent policy. All customers of its mass spectrometers were required to sign a contract whereby patent rights for any improvements to the instruments would be assigned to CEC. In fact, the early CEC users' meetings provided the forum through which improvements were fed back to the company and incorporated into new and more advanced mass spectrometers.

Early in 1952 a group of mass spectrometry practitioners gathered at the Philadelphia headquarters of the American Society for Testing and Materials (ASTM) to decide whether a separate organization dedicated solely to the field of analytical mass spectrometry should be established. A steering committee was formed to study the recommendation, but no formal action was taken until the annual meeting of the Pittsburgh Conference on Analytical Chemistry and Applied Spectroscopy (PITTCON). Leading the effort at PITTCON was Frederick Mohler, a respected analytical chemist who headed mass spectrometry research at the National Bureau of Standards in Washington, D.C. In addition to organizing a separate mass spectrometry symposium of about one hundred participants at the PITTCON meeting, Mohler polled the participants in an effort to establish a new independent organization. The vote carried, and permanent officers were elected. The First Annual Conference on Mass Spectrometry and Allied Topics was held in Pittsburgh in 1953, in conjunction with PITTCON. Subsequent conferences were held independently of PITTCON at locations throughout the country and attracted mass spectrometry experts from abroad, especially Europe and the Pacific Rim.

Under the aegis of ASTM Committee E-14, annual conferences were held to exchange information and research results on the latest developments in analytical mass spectrometry.

MASS SPECTROMETRY	1996

Studies of Boron isotope ratios of meteorites and lunar rocks imply that the isotopic composition of boron in the solar system is homogeneous.

HISTORY	1996

In the Hague, war crimes trials begin for those accused of atrocities in the Balkan wars.

Guatemala's thirty-five-year civil war ends.

Mad cow disease plagues British livestock.

The Conference That Almost Wasn't

In 1975 the ASMS annual conference had been scheduled to take place at the Rice Hotel in Houston, Texas. The hotel, named after William Marsh Rice, the founder of Rice University, was at the time the most prestigious venue in the Houston area, with a long and colorful local history dating back to the 1880s. Conference arrangements were usually made several years in advance. In any case the ASMS board discovered, quite by accident, that the Rice Hotel was going to close in fall 1974. This situation created an immediate panic. The annual conference attendance was in the neighborhood of six to seven hundred scientists, most of whom were planning on meeting in Houston. It was much too late to start shopping for hotels with the conference barely six months away. No venues were available for the scheduled time of the conference in the Houston area.

Although the exact details have been lost to posterity, somehow the board was able to secure an alternate location at the Houston Shamrock Hilton. The catch was that the hotel was only available for the week containing the Memorial Day holiday. While there was some grumbling from attendees about having to give up their holiday, members understood the precariousness of the situation and attended nevertheless. However, this last-minute schedule adjustment set a precedent of having the annual conference during the week containing the Memorial Day holiday. Subsequent boards of directors realized that the week was hard for hotels to sell. Therefore the ASMS vice president for arrangements was able to negotiate better hotel rates and meeting-space terms if the conference was scheduled for that week. Interestingly, the Rice Hotel, shuttered for twenty years, reopened as luxury apartments in 1997. While the sudden change of venue in 1975 did not have an adverse affect on the ASMS annual conference, it would be hard to imagine the damage such an event would cause today, with over 3,500 attendees coming from all over the world.

The presentations took on a decidedly applied flavor, focusing on the solution of complicated analytical problems, such as the study of petroleum products, mixture analysis, instrumentation, and process control. Consider, for example, the institutional affiliations of the twenty-six speakers who participated in the 1953 conference. Only three speakers were from academic institutions. The rest of the program was filled with presentations by analytical chemists employed in government and industrial laboratories. Participating industrial firms included Imperial Chemical Industries, Standard Oil of Indiana, Westinghouse Electric, RCA, DuPont, General Electric, Metropolitan-Vickers, Humble Oil, and Dow Chemical. These early meetings also harbored a group with a small but growing interest in fundamental and theoretical topics. At the 1953 meeting, for example, participants listened to presentations on the physics and ion chemistry responsible for the formation of charged particles under conditions of electron ionization.

As mass spectrometry moved from the petroleum research laboratory and the process refinery into new fields of scientific research, an increasing number of practitioners attended the Committee E-14 conferences. Expanded programs and the appearance of multiple parallel sessions placed serious strains on the community's ability to exchange ideas. The organizational response to this problem first appeared in the 1970s, when poster sessions were

1997

	Unit resolution by Fourier transform ion-cyclotron resonance mass spectrometry demonstrated at 112kDa.	The first ASMS Biemann Medal is awarded.

1997

Hong Kong is returned to Chinese control.	Pathfinder is the first moving vehicle to land on Mars.	In Zaire, Joseph Mobutu is overthrown and the country is named Congo once again.

The Community of Mass Spectrometrists

introduced. Although not as formal as the oral presentations, this new venue did stimulate productive interaction among the conference participants.

The American Society for Mass Spectrometry

As ASTM Committee E-14 continued to grow, more participants at the annual conferences began to focus their presentations on the fundamental science behind mass spectrometry. While ASTM administration in Philadelphia seemed noncommittal about the evolution of Committee E-14 into a first-rate scientific conference, there was a feeling among some members that the best interests of attendees and even of ASTM were not being served. Subcommittees of E-14, established to prepare standards for the use of mass spectrometry in the analysis of chemicals, received less attention as attendance grew. The membership, which originally comprised primarily industrial chemists from petroleum and chemical companies and some government laboratories, had shifted by the mid-1960s to a group of scientists with affiliations in other branches of academic chemistry. Consequently, in 1969, the tension between the academic and industrial groups became so contentious that a vote was held to consider the creation of a new organization separate from ASTM. The outcome was the emergence later that year of the American Society for Mass Spectrometry (ASMS).

In 1970 the slate of ASMS officers organized and ran the conference with ASTM activities proceeding as a committee of ASMS. In deference to the important role that ASTM E-14 had played in the evolution of the new organization, the chairman of ASTM E-14

Figure 1. Attendance at the Annual Meetings on Mass Spectrometry and Allied Topics.

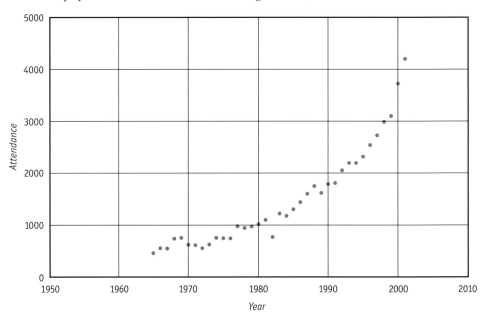

MASS SPECTROMETRY   1998

Fast chemical reactions are studied by femtosecond pump-probe spectroscopy at Penn State University.

HISTORY   1997

An economic crisis begins in Thailand and spreads through eastern Asia.

Dolly the sheep is born, the first large mammal cloned from an adult specimen.

1998

President Bill Clinton is impeached by the U.S. House of Representatives but is later acquitted by the Senate.

was ex-officio a member of the board of directors of ASMS. Thus, the scientific content presented at the annual conference was handled by ASMS, and the development of testing and measurement standards remained the sole responsibility of ASTM Committee E-14. This arrangement continued until 1986, when the ASMS board recommended that its formal ties with ASTM be dropped.

From the beginning ASMS experienced steady growth in membership and attendance at the annual conference, as shown in Figure 1. This expansion indicated that more scientists were using mass spectrometry in an increasingly diverse number of fields, but as the

Award for a Distinguished Contribution in Mass Spectrometry

1990 **Ronald D. Macfarlane**
Plasma desorption ionization

1991 **Michael Barber**
Fast atom bombardment

1992 **John B. Fenn**
Electrospray ionization

1993 **Christie G. Enke and Richard A. Yost**
Triple quadrupole mass spectrometer

1994 **Donald F. Hunt**
Negative ion chemical ionization

1995 **Keith R. Jennings**
Discovery and application of collision-induced dissociation

1996 **Frank H. Field and Burnaby Munson**
Development and application of chemical ionization mass spectrometry

1997 **Franz Hillenkamp and Michael Karas**
Discovery of matrix-assisted laser desorption ionization mass spectrometry

1998 **David A. Dahl and Don C. McGilvery**
SIMION ion optics design and analysis programs

1999 **Melvin Comisarow and Alan G. Marshall**
Fourier transform ion-cyclotron resonance mass spectrometry

2000 **Boris A. Mamyrin**
Reflectron time-of-flight mass spectrometry

2001 **George C. Stafford**
Developments in quadrupole ion trap mass spectrometry

2002 **William J. Henzel, John T. Stults, and Colin Watanabe**
Peptide mass fingerprinting for proteomics

1999
Noncovalent associations are preserved on a MALDI surface without immobilization.

1999

India and Pakistan test their first nuclear weapons.

Suharto steps down as dictator of Indonesia.

Violence tears through the Indonesian province of East Timor after a pro-independence vote.

"If the Shoebox Fits . . . "

Early organizers of the ASMS annual conference decided that there would not be an instrument exposition in conjunction with the conference. The mass spectrometer instrument companies liked the concept of marketing at ASMS conferences by means of hospitality suites. It was much more cost-effective and less stressful to provide refreshments at the end of the day than to move a complex instrument to the meeting site, set it up, and work long hours to overcome the inevitable bugs that would pop up just when you were ready to demonstrate a special feature to the customer who was about to sign on the dotted line. However, with time, corporate sponsors began to include companies that did not sell large instruments. They sold much smaller items, such as specialized glassware, books, software, and reagents. In addition, some of the large instrument companies wanted to show their smaller mass spectrometers. These corporate sponsors wanted very much to bring along a selection of their wares to have at their hospitality suites for prospective customers to see firsthand. The board of directors was in the middle of the competing interests of the two groups. The board sided with those who did not want to have an instrument exposition, since it would result in additional work for board members as well as additional costs to the society. Since there were already a number of exposition venues that satisfied the needs of both the scientific community and the corporate sponsors, the board established and held firm on a policy that no mass spectrometers were to be displayed at the annual conference.

However, the corporate sponsors with small products finally won a concession from the opposing interests. The board decided that any item that could be "hand carried" could be exhibited at the conference. One of the mass spectrometer manufacturers was able to hand carry a small mass spectrometer into its suite in 1968, so the rule had to be changed. The board then settled on the "shoebox" rule. Anything smaller than a shoebox could be brought to the conference and shown in the sponsor's hospitality suite. The shoebox rule was enforced for a number of years, but with the development of even smaller mass spectrometers it was not adequate to the task of preventing the exposition of equipment. The shoebox rule eventually gave way to the "no pressure, no vacuum" rule for the exposition of equipment. This kept out chromatographs and mass spectrometers, but computers and computer software were admissible.

However, the slide down the slippery slope continued, and in 2000 at the Long Beach, California, conference, the periphery of the poster session area was set up with corporate booths for exposition. These booths provided an alternative to hospitality suites and a more substantial venue for manufacturers than either the corporate or scientific posters. The ASMS board is simply facing the reality that corporate membership in the past was composed primarily of mass spectrometry manufacturers. With the rapid increase in corporate members in related technologies there was an increased demand for hospitality suites that exceeded their availability. In addition, many manufacturers found their suites to be located too far from the evening traffic, or they simply could not compete well for visitors. Thus, the current policy is to provide a variety of alternative venues that suits both larger and smaller companies.

attendance increased, the annual conference ran the risk of becoming more impersonal. ASMS was eminently successful in its primary goal of promoting mass spectrometry and broadening the scope of its applications. While the organization's officers did not want the conference to lose its friendly, intimate ambience, they could not deny attendance to their new, younger colleagues who also were interested in mass spectrometry.

MASS SPECTROMETRY	1999	2000
	Higher fullerenes are discovered in carbonaceous chondrites.	A noncovalently associated macromolecule greater than one megadalton is ionized and detected by electrospray ionization time-of-flight mass spectrometry.

HISTORY	1999		2000
	Brian Jones and Bertrand Piccard travel nonstop around the world by balloon.	An earthquake kills 17,000 people in Turkey.	Slobodan Milosevic is overthrown as the leader of Yugoslavia.

Biemann Medal

The Biemann Medal recognizes a significant achievement in basic or applied mass spectrometry made by an individual early in his or her career. The award is presented in honor of Professor Klaus Biemann, of MIT, and is endowed by contributions from his students, postdoctoral associates, and friends.

1997 **Scott A. McLuckey**
Studies of multiply charged ions

1998 **Robert R. Squires**
Studies in gas-phase ion chemistry

1999 **Matthias Mann**
Mass spectrometric applications in protein chemistry and molecular biology

2000 **Julie A. Leary**
Structural analysis of carbohydrates by mass spectrometry

2001 **Peter B. Armentrout**
Mass spectrometry of metal ion complexes of biological and environmental significance

2002 **Ruedi Aebersold**
Isotope-coded affinity tags and related work in proteomics

Intense interest in the recently introduced fast atom bombardment (FAB) ionization technique by researchers wanting to analyze biological compounds precipitated a two-day conference on the subject in 1980. At about the same time and independently of ASMS a conference on tandem mass spectrometry was organized in Monterey, California. During its early years this meeting, known as the Asilomar Conference on Mass Spectrometry, was considered to be a potential competitor to the ASMS annual conferences. In 1987, however, the Asilomar conference formally affiliated with ASMS. Six years later, in 1993, the conference became the first of the regularly scheduled, multi-day meetings held by ASMS. Similarly, in 1988, the first of a series of highly focused meetings on specialized topics at Sanibel Island off the Florida coast was held. These smaller, more specialized meetings have preserved the more intimate format and ambience of the earlier conferences.

In the early days of analytical mass spectrometry no formal university courses existed to teach the subject to students interested in entering the field. Most practitioners were trained through hands-on experience. While graduate programs specializing in mass spectrometry did appear over time, many researchers in the field even today have not benefited from formal training. To some extent this absence of formalized training reflects the transformation of mass spectrometry instrumentation itself. No longer is it necessary for the practitioner to

---

2001

Moderate parties win parliamentary elections in Iran.

Vicente Fox becomes president of Mexico in the country's first legitimate democratic elections.

The sequencing of the human genome is completed.

The first completely self-contained artificial heart is implanted in Robert L. Tools, who lives for several months on the device.

Chairmen of the ASTM Committee E-14 on Mass Spectrometry

1953–1954	W. S. Young	1961–1962	V. H. Dibeler
1955–1956	M. J. O'Neal	1963–1964	R. E. Fox
1957–1958	W. Priestley, Jr.	1965–1966	N. D. Coggeshall
1959–1960	R. A. Friedel	1967–1968	H. M. Rosenstock

Presidents of the American Society for Mass Spectrometry

1969–1970	J. L. Franklin	1986–1988	Gerry Meisels
1970–1972	Richard Honig	1988–1990	Ronald A. Hites
1972–1974	Frank H. Field	1990–1992	Robert C. Murphy
1974–1976	Harry J. Svec	1992–1994	Henry M. Fales
1976–1978	Jean H. Futrell	1994–1996	Christie G. Enke
1978–1980	James A. McCloskey	1996–1998	Veronica M. Bierbaum
1980–1982	Burnaby Munson	1998–2000	Robert J. Cotter
1982–1984	Catherine C. Fenselau	2000–2002	Richard M. Caprioli
1984–1986	R. Graham Cooks	2002–2004	Catherine E. Costello

be a skilled craftsman intimately familiar with every detail of the instrument's design and operation. Today analyses can be run on bench-top mass spectrometers, essentially "black boxes" operated by laboratory technicians. The mass spectrometer has become a routine and standardized analytical tool.

Even so, training and continuous education has remained an important professional activity within the mass spectrometry community. As we have seen, instrument manufacturers, such as CEC and General Electric, helped train their customers through early users' meetings. The American Society for Mass Spectrometry has taken a more proactive role in continuing education. Beginning in the late 1970s, ASMS organized a series of short courses typically held before the annual conference. Qualified ASMS members with hands-on experience were selected as instructors for these courses, and enrollment was limited. Initially, the courses centered on interpretation of mass spectra and quantitative analysis. At the 2001 annual conference in Chicago ten short courses were offered to participants. Topics ranged from characterization of proteins and peptides to combinatorial chemistry and practical liquid chromatography mass spectrometry.

Specialized conferences, annual meetings, and advanced training courses enable scientific communities to distinguish themselves from other groups. Also crucial to this professional identity is the publication of conference proceedings and the distribution of refereed journals. The early users' meetings sponsored by CEC did not possess the resources necessary to sup-

An energy crisis in California leads to scheduled rolling blackouts in some parts of the state.

Former Yugoslavian ruler Slobodan Milosevic is handed over to a United Nations tribunal by Yugoslavian authorities to stand trial for genocide and other crimes against humanity allegedly committed during the Balkan wars of the 1990s.

port such activities. The American Society for Testing and Materials, however, began publishing its E-14 Committee proceedings in 1961. This arrangement remained in effect until ASTM and ASMS parted company in the mid-1980s. By the early 1990s ASMS was turning out thick conference proceedings containing nearly two thousand pages of text. The adoption of CD-ROM technology by the organization later in the decade alleviated to a large extent the content overload commonly experienced by members trying to wade through reams of text. Another outlet for information was the creation of the *Journal of the American Society for Mass Spectrometry*. In 1990, the first year of publication, *JASMS* (or "Jazz-Mass," as it is informally called) numbered six issues with eight to ten articles per issue. Today, *JASMS* has become, along with an electronic version, one of the most frequently cited journals in the chemical literature.

The dawn of the twenty-first century is a particularly appropriate time to reflect on the history of the mass spectrometry community and the development of the ASMS. During the last one hundred years, the field has transformed itself from a small subdiscipline of academic physics into one of the largest and most diverse fields of analytical chemistry. The growth in numbers of members and quality of scientific presentations at the ASMS annual conferences could not have occurred without the dedicated efforts of many individual scientists and engineers over the past fifty years. Although the names of many of these people are unknown to us today, each made contributions upon which others could build. Some are still with us and continue to give their time and intellectual resources to help guide ASMS. We can only hope that the vision, insight, and willingness to contribute time and energy to the goal of "promoting and disseminating knowledge of mass spectrometry and allied topics" will come forth from present and future members of ASMS so that the Annual Conference on Mass Spectrometry and Allied Topics will continue to grow and shape the future of this dynamic and important field of chemistry.

2002

The fiftieth Annual Conference on Mass Spectrometry and Allied Topics is held in Orlando, Florida.

Outbreaks of foot-and-mouth disease devastate livestock in Britain.

Terrorists crash jetliners into the World Trade Center in New York City, the Pentagon, and a field in Shanksville, Pennsylvania.

Photo Credits

Chapter 1

Page 2: Cavendish Laboratory, University of Cambridge. Page 3: (left) by permission of the Syndics of Cambridge University Library. Photo supplied by AIP Emilio Segrè Visual Archives; (center) The Fisher Collection, Chemical Heritage Foundation; (right) Edgar Fahs Smith Collection, University of Pennsylvania Library; (timeline, left to right) AIP Emilio Segrè Visual Archives, Landé Collection; © Bettmann/CORBIS; Edgar Fahs Smith Collection, University of Pennsylvania Library. Page 4: (top) © Bettmann/CORBIS; (bottom) photogravure by Gen. Stab. Lit. Anst., AIP Emilio Segrè Visual Archives, Weber and E. Scott Barr Collections; (timeline) Edgar Fahs Smith Collection, University of Pennsylvania Library. Page 5: (timeline) Hulton Archive by Getty Images. Page 6: (top left) Edgar Fahs Smith Collection, University of Pennsylvania Library; (top right) Cavendish Laboratory, University of Cambridge; (bottom) Cavendish Laboratory, University of Cambridge; (timeline) Edgar Fahs Smith Collection, University of Pennsylvania Library. Page 7: © The Nobel Foundation. Page 8: © The Nobel Foundation. Page 9: (timeline) National Archives and Records Administration. Page 10: (top) AIP Emilio Segrè Visual Archives; The Burndy Library, Dibner Institute for the History of Science and Technology, Cambridge, Massachusetts; (bottom) Special Collections, Research Center, University of Chicago Library. Page 11: University of Minnesota Archives. Page 12: (top) University of Minnesota Archives; (bottom) Harvard University Archives. Page 13: University of Minnesota Archives; (timeline) National Archives and Records Administration. Page 14: A. O. Nier et al., "Nuclear Fission of Separated Uranium Isotopes," *Physical Review* 57 (15 Mar. 1940), 546. © 1940 American Physical Society. Page 15: Bancroft Library, University of California, Berkeley (Lawrence, EO–POR 75). Page 16: National Archives and Records Administration; (timeline) National Archives and Records Administration. Page 17: National Archives and Records Administration. Page 18: Schenectady Museum.

Chapter 2

Page 20: Bildarchiv der Österreichischen National-bibliothek. Page 21: Austrian Academy of Sciences. Page 22: (timeline, both) National Archives and Records Administration. Page 23: (top) 3M; (bottom) Case Western University, School of Medicine. Page 27: © The Nobel Foundation. Page 29: (timeline) National Archives and Records Administration.

Chapter 3

Page 32: National Archives and Records Administration. Page 33: National Archives and Records Administration; (timeline) Edgar Fahs Smith Collection, University of Pennsylvania Library. Page 39: (left) photograph by Neal Douglas. Photograph provided by Frank H. Field; (right) © Bettmann/CORBIS. Page 40: (left) photograph taken at the Defense

Standards Laboratories (DSL), Maribyrnong, Australia, 1962. Photograph provided by John Occolowitz; (right) Schenectady Museum; (timeline) Hulton Archive by Getty Images. Page 41: (top left) by permission of DuPont; (bottom left) reprinted with permission from Honeywell International Inc. Bendix is a registered trademark of Honeywell. (right) Chemical Heritage Foundation Image Archives, Othmer Library of Chemical History. Page 43: (timeline) National Archives and Records Administration.

Chapter 4

Page 46: Special Collections Department, J. Willard Marriott Library, University of Utah. Page 49: (timeline) National Archives and Records Administration. Page 50: Department of Chemistry, State University of New York, Stony Brook. Illustration by Joseph W. Lauher. Page 53: photograph by Orren Jack Turner. AIP Emilio Segrè Visual Archives.

Chapter 5

Page 56: © A. Barrington Brown, Photo Researchers, Inc.; courtesy James Dewey Watson. Page 57: National Institutes of Health Web site. Page 59: (left) Columbia University Archives—Columbiana Library; (right) H. C. Urey, G. M. Murphy, and F. G. Brickwedde, "A Hydrogen Isotope of Mass 2 and Its Concentration," *Physical Review* 40 (1 April 1932), 1. © 1940 American Physical Society. Page 60: (left) Columbia University Archives—Columbiana Library; (right) © The Nobel Foundation; (timeline) National Archives and Records Administration. Page 63: (timeline) National Archives and Records Administration. Page 65: (timeline) National Archives and Records Administration. Page 66: (timeline) National Archives and Records Administration. Page 67: (timeline) Hulton Archive by Getty Images. Page 68: (timeline) Hulton Archive by Getty Images.

Chapter 6

Page 72: © CORBIS. Page 73: (timeline) National Archives and Records Administration. Page 74: © Bettmann/CORBIS; (timeline) © Bettmann/CORBIS. Page 75: (timeline) © Bettmann/CORBIS. Page 78: (timeline) Hulton Archive by Getty Images. Page 79: (timeline) National Archives and Records Administration. Page 81: (timeline) © Bettmann/ CORBIS. Page 85: (timeline) National Archives and Records Administration.

Chapter 7

Page 88: NASA. Page 91: NASA; (timeline, left to right) National Archives and Records Administration, Abbie Rowe, National Park Service/John F. Kennedy Library. Page 92: (timeline) National Archives and Records Administration. Page 94: NASA. Page 95: NASA. Page 96: NASA. Page 97: artwork by Paul Hudson, NASA; (timeline) National Archives and Records Administration. Page 98: (top) NASA; (bottom) University of Minnesota Archives. Page 99: (both) NASA; (timeline) National Archives and Records

Administration. Page 101: (left) NASA; (right) Hulton Archive by Getty Images. Page 102: (timeline) National Archives and Records Administration.

Chapter 8

Page 104: photograph by Gene Daniels. National Archives and Records Administration. Page 105: U.S. Patent and Trademark Office Web site; (timeline) National Archives and Records Administration. Page 106: by permission of Rachel Carson Council, Inc.; photograph by Edwin Gray © 1951. Page 107: photograph by Gene Daniels. National Archives and Records Administration. Page 109: (top left) Fermi National Accelerator Laboratory; AIP Emilio Segrè Visual Archives, Physics Today Collection; (top right and bottom) ThermoFinnigan; (timeline) Hulton Archive by Getty Images. Page 110: photograph provided by Michael Story; (timeline) Reuters/Stringer/Hulton Archive by Getty Images. Page 111: (left) photograph by Michael Story; (right) ThermoFinnigan. Page 112: (all) Agilent Technologies, Palo Alto, California. Page 116: R E. Finnigan, D. W. Hoyt, and D. E. Smith, "Priority Pollutants II—Cost-Effective Analysis," *Environmental Science & Technology* 13 (May 1979), 534. © Robert E. Finnigan; (timeline) Ronald Reagan Library. Page 117: United States Environmental Protection Agency, Office of Emergency and Remedial Response. Page 118: George Bush Presidential Library.

Chapter 9

Page 120: © Allsport. Page 122: (timeline) Reuters/Stringer/ Hulton Archive by Getty Images. Page 125: © Joe McDonald/CORBIS; (timeline, left to right) Reuters/Mike Blake, UPI/Hulton Archive by Getty Images; Reuters/David Brauchli/Hulton Archive by Getty Images. Page 126: © Jean-Pierre Lescourret/ CORBIS. Page 128: © Layne Kennedy/CORBIS. Page 129: © AFP/CORBIS. Page 130: (timeline) Express Newspapers/7046/Hulton Archive by Getty Images.

Chapter 10

Page 132: American Society for Mass Spectrometry, donated by A. O. Nier. Page 136: (timeline) Reuters/ Roslin Institute by Getty Images.

Index

Page references for photos are in italic type.